Acid Rain

A Plague Upon the Waters

by Robert Ostmann Jr.

DILLON PRESS, INC., MINNEAPOLIS, MINNESOTA

To P.G.

Library of Congress Cataloging in Publication Data

Ostmann, Robert.
  Acid rain.

  Includes bibliographical references and index.
  1. Acid rain.   I. Title
TD196.A25077   363.7'394   82-7443.
ISBN 0-87518-224-0   AACR2

Acknowledgments

I would like to acknowledge all those who lent me support and assistance during the years I spent researching and writing this book. I would especially like to thank Steve Isaacs, the inimitable editor of the late *Minneapolis Star,* who cheered me on and allowed me to take up huge chunks of memory in the newspaper's computer system to put this book together.

CONTENTS

1. *A Threat to Planet Earth*

I find refuge in a Long Island phone booth and watch the rain sweep in a great curtain across the empty street. In a clearing in the Colorado Rockies, I watch a young woman climb a wooden tower to collect a sample of the rain. I stand in the parking lot of the Burbank airport, looking at the sky, wondering if today the rain will bring down the thick southern California haze. I kneel by a high mountain lake in the heart of the wild Adirondacks and dip my hand in water where, because of the rain, fish can no longer survive.

The rain is the reason I've come to each of these places and to many others during the past two years. But not the pure, energizing rain we've always taken for granted; the cleansing, life-giving rain that the prophet Isaiah said, "comes down and waters the earth and makes it bring forth and bud;"[1] the rain of Mohammed, "water from heaven [that] causes the earth to revive."[2]

I am here because something has gone terribly wrong with the rain in New York, Ohio, Minnesota, Colorado, California, and in countless other places, over hundreds of thousands of square miles of low plains and high peaks in North America, Europe, Asia, even the Arctic.

The rain has turned to acid.

Sulfuric showers. April acid. Nitric rainbow. These are nasty, toxic, corrosive laboratory names for what was once the universal symbol of natural purity.

We spent a decade and more in the 1960s and 1970s calling for a halt to the ever-increasing pollution of the planet, but we somehow forgot about the rain. We turned a vigilant eye to pipes disgorging vile fluids and to smokestacks soiling neighborhoods with clouds of noxious emissions. But when the air appeared cleaner, most of us looked no farther. We just assumed that of all things, the rain would remain pure. We did not realize until the raindrops were already sour that the destruction had begun; that if you pump great volumes of poison into the complex global organism of the environment, the ill effects will spread far beyond the point of injection.

After the first alarms from scientists—largely ignored—led to more and louder alerts, we finally began to refocus our sights. What we found is that the poison had indeed spread, insidiously, on a scale we had never imagined. We found that both the rain and the snow actually had become dilute sulfuric, nitric, and in some cases hydrochloric acid. The transformation is the result of a complicated, and not yet fully deciphered, atmospheric recipe whose key ingredients are sun, wind, water, and chemical pollutants. These pollutants— mainly sulfur dioxide and nitrogen oxides—are released into the atmosphere wherever coal and, to a lesser degree, oil or natural gas are burned: from the smokestacks of electric generating plants, metal smelters and industrial boilers, and from the exhaust pipes of motor vehicles. The same pollutants also arise from natural sources such as volcanic eruptions, forest fires, and the slow bacterial decomposition of organic matter. The rapidly increasing portion contributed by humans, however, has caused the trouble. In one year the sulfur dioxide emissions from a large coal-fired power plant can equal the huge amount released by the May 18, 1980, eruption of Mount St. Helens.

Each year human activity injects at least 100 million metric tons of sulfur dioxide and 35 million metric tons of nitrogen oxides into an already polluted atmosphere.[3] Americans contribute some 30 million metric tons of sulfur dioxide and about 26 million metric tons of nitrogen oxides to the acid-forming pollutants over North America.[4] The Canadians, primarily their metal smelting industry, release another 5 million metric tons of sulfur dioxide and 2 million metric tons of nitrogen oxides.[5] Over the rest of the Northern Hemisphere, the European nations are responsible for more than 50 million metric tons of sulfur dioxide and 4 million metric tons of nitrogen compounds.[6] Then there are the Russians, whose current emissions are unknown, and, increasingly, the Chinese, who with sulfur dioxide emissions of 15 million metric tons, are rushing headlong into a highly industrialized, acidic future.[7]

The Earth's weather systems weave together these innumerable pollution plumes into great regional masses of contaminated atmosphere that mix, swirl, and carry the sulfur and nitrogen

Global Human-Caused Sulfur Dioxide Emissions

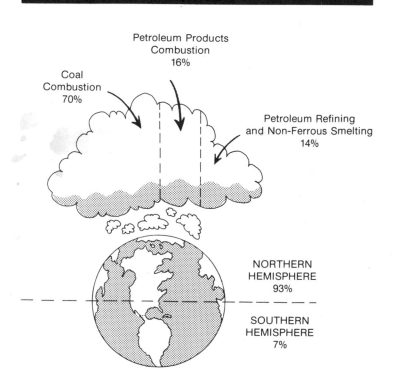

Petroleum Products Combustion 16%

Coal Combustion 70%

Petroleum Refining and Non-Ferrous Smelting 14%

NORTHERN HEMISPHERE 93%

SOUTHERN HEMISPHERE 7%

compounds hundreds, and sometimes thousands, of miles from where they were originally discharged into the air. Scientists now believe that from one-half to two-thirds of the compounds that enter these masses of atmospheric pollution react with moisture and other particles in the air and are transformed into molecules of dilute sulfuric and nitric acid that condense and fall to earth as acid rain or snow. In areas that receive little rainfall, most acid-forming pollution reaches the earth as dry particles—a phenomenon known as dry deposition—or combines with water particles to form acid dew or mist. Southern California has its own special brand of acid smog.

Taken together in all of its forms, wet and dry, acid deposition has emerged in recent years as one of the most serious environmental threats to life on our planet. In fact, its visible and devastating ecological effects have caused scientists and governments around the world to sound cries of alarm. In the scientific community today, there is general—though not unanimous—agreement that acid deposition has begun to extinguish entire species of fish and other aquatic life forms in vast areas of the Northern Hemisphere. In addition, many scientists believe that continuing or worsening acid deposition could reduce the productivity of vital forests and farmlands, disrupt the crucial, life-sustaining process of plant photosynthesis in large areas, and poison some drinking water supplies and food fish stocks.

Acid has also been linked to the corrosion of billions of dollars worth of buildings, cars, art-works, and other human-made objects—at least $5 billion each year in damage in the United States alone, and hundreds of millions more in Canada and Europe.

Finally, even in places that for reasons of geology or meteorology might escape some or all of these deleterious effects of acid deposition, recent studies have shown that some people who are susceptible to heart and lung diseases will die because they breathe air contaminated with acid particles.

Over the course of a year, the rain and snow falling on large sections of North America and northern Europe are now from five to forty times more acidic than they would be if humans were not burning fossil fuels.[8] Individual storms in these areas frequently produce precipitation that is hundreds to thousands of times more acidic than normal. In Wheeling, West Virginia, a 1978 storm was recorded with rainfall about five thousand times the acidity of normal rain—the equivalent of concentrated lemon juice in acid strength.[9] In much of the eastern United States, southeastern Canada, and western Europe, the average rain and snow is about the acid equivalent of orange juice. The Great Lakes basin of North America is especially hard hit because almost all watersheds draining into the lakes are now receiving acid precipitation.

Two of the most disturbing aspects of the acid rain phenomenon are its increasing severity and its ever-widening scope. In 1955 acid precipitation was recorded in twelve states in the northeastern United States. By 1972 the area of the nation experiencing acid precipitation had spread dramatically to the south and west to include all of eastern North America (except for the southern tip of Florida and northern Canada), the Rocky Mountains, and the West Coast.[10] In fact, the rainfall in Georgia and Florida today is about as acidic as the rainfall was in the most seriously affected parts of the Northeast twenty-five years ago.[11] A 1981 National Wildlife Federation analysis of acid precipitation patterns and geology found that fifteen states east of the Mississippi River are extremely vulnerable to the harmful effects of acid rain, ten others face moderate threats, and "it can't be long before every state in the union is affected."[12]

Far more disturbing, however, is the apparent global spread of acid precipitation. Noted in northern Europe and the northeastern United States twenty-five years ago, acid rain and snow are now found throughout the Northern Hemisphere, including the Arctic, and recently acid rain has been discovered in South America and Asia.

Acid rain respects no national border. Germany's pollution becomes Scandinavia's acid, and therein lies the most difficult barrier to any timely remedy to the acid rain problem. The Swedes and

Acid Sensitive Areas of North America

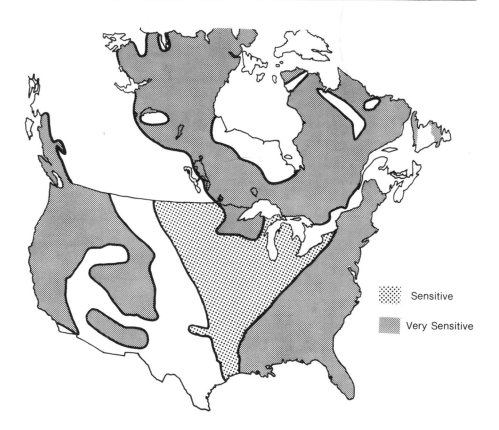

Sensitive

Very Sensitive

Norwegians have been complaining loudly about the acid falling on their countries for more than a decade. Despite expressions of willingness from the offending nations to look for a solution, the acid-forming pollutants still waft skyward.

And if the acid precipitation problem has not been made difficult enough by its severity and its scope, then it is made maddeningly vexing by its timing. For it has come to attention at the same time that the United States and all other industrialized nations are under pressure to shore up their sagging oil-based economies by burning more coal—the source of the acid rain crisis.

What happens when the airborne acid hits the earth? That depends, to some extent, on how strong the acid is and where it falls. Acid rain can cause immediate damage wherever it lands. Plant tissues can be burned. The marble or limestone on buildings can crumble into a fragile crust. The Statue of Liberty, the Washington Monument, the Parthenon, and many other historic structures around the world are being corroded by acid pollution. Metal—even stainless steel—can be corroded by acid. Paint, as U.S. paint manufacturers have noted in recent years, can be worn and spotted by exposure to acid rains. And if the acid is bound up in a fog or mist, human lung tissue can be scarred, and respiratory illnesses can be aggravated fatally.

From an ecological point of view, the insidious nature of the acid rain threat lies in the slow, gradual, subtle environmental degradation resulting from long exposures to small amounts of acid. The areas of the world that are most vulnerable to drenchings of dilute sulfuric and nitric acid are those not blessed with abundant quantities of acid-neutralizing minerals in the soil. Acid falling on these areas accumulates in the soil and in the waters, gradually using up whatever chemical protection nature has provided and eventually bringing about a drastic alteration in the chemical and biological environment.

By cruel coincidence, the areas most susceptible to the effects of these continual, dilute acid rains are among the most wild, splendid, and economically valuable regions of the world.

In Scandinavia increasingly acidic precipitation since the early 1900s has decimated the important commercial salmon fishing industry in sections of southern Norway and Sweden. No longer will the majestic salmon swim up the acidified rivers and streams to spawn.[13]

In the Adirondacks, the vast mountain sanctuary in upstate New York, acid rain has killed the fish in about 200 of the region's best sport fishing lakes. Many more lakes are on the way to lifelessness under year after year of falling acid.[14]

In southeastern Canada, clouds of pollutants from the United States mix with those of the Inco Ltd. copper-nickel smelter, one of the world's single most expulsive sources of acid-forming pollution. Frantic environmental officials have estimated that 50,000 lakes—lakes that provide food and recreation for hundreds of thousands of people—will succumb to the effects of acidification before the end of this century.[15]

In the Boundary Waters Canoe Area of northeastern Minnesota—the largest wilderness area in the eastern United States—recent studies have shown that annual rainfall is now at the threshold of acidity at which damage to aquatic life begins to appear in lakes and streams. The studies have concluded that as many as 2,500 lakes in the region could become acidified.[16]

The list goes on and on. Isle Royale, an isolated jewel of a national park in the middle of Lake Superior, is being subjected to acid rain and snow. Almost all of the national parks in the so-called "Golden Crescent" in the southwestern United States—Bryce Canyon, Zion, Canyonlands—are in the path of acid fallout from coal-fired power plants that are springing up throughout this rapidly developing region. The snows and rains in the Colorado Rockies have become increasingly acidic during the last four years. The mists of the Great Smoky Mountains in the East and the Sierra Nevadas in the west have an acid tinge and a North Carolina newspaper, the *Charlotte Observer*, reports the acidity of the rain on its weather page.

Wherever it falls, acid works its harm on living things by altering the chemical composition of their environment. Most plants, animals, and microorganisms have evolved to exist in a defined

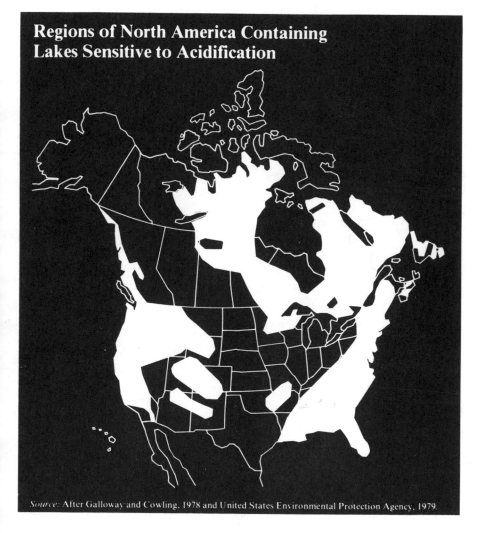

Regions of North America Containing Lakes Sensitive to Acidification

Source: After Galloway and Cowling, 1978 and United States Environmental Protection Agency, 1979.

niche of acidity. Just as human tissues are damaged by contact with battery acid, other living systems are vulnerable to other forms of acid encounters. Scientists already know a great deal about how acidification can disrupt the natural order from the level of the molecule to the scale of whole populations:

● Enzymes, which catalyze vital reactions inside cells, are dependent on the acidity of the surrounding environment and are rendered less effective or totally inactive by increases in acid levels.

● Proteins, which comprise a significant part of the matter in all cells, undergo changes in geometry and function when acidity is altered.

● Organisms generally cannot reproduce and maintain themselves in optimal fashion unless their environment remains within a fixed range of acidity.

● Acidification of an environment frees up toxic metals such as aluminum, mercury, and lead that would otherwise remain safely bound.

● Acidification limits the diversity of an eco-system by preventing the establishment or flourishing of acid-sensitive species.

To date the most dramatic effects of acid precipitation have occurred in freshwater lakes and streams. Acid and most water creatures, large and small, just do not mix. In the Adirondacks, Scandinavia, and Canada, scientists have found that when the acidity of lakes and streams exceeds a certain threshold, some of the youngest fish die, and the rest cease reproducing. Then, as each generation of older fish dies and is not replaced, the stock of fish in a particular body of water is progressively depleted until no fish remain.

The downpouring of acid has damaged more than just fish. Frogs and salamanders are unable to breed in acidified pools; plant tissues in the leaves of soybeans and other crops are burnt by acid; and basic life processes such as photosynthesis, nitrogen fixation, and the decay of organic matter can be disrupted or prevented by acidification.

One effect of acidification, the toxic metal liberating capacity of acid rain, poses what may be the most immediate threat to human welfare. In areas subjected to acid rains but still supporting aquatic life—such as lakes in northern Minnesota near Voyageurs National Park and the Boundary Waters Canoe Area Wilderness—high levels of mercury have been found. These levels in fish living far from any industrial discharge of the poisonous metal have been explained by some scientists as the accumulation of mercury liberated by acid precipitation. If their theory is correct, acidification could pose a direct threat to those who regularly consume fish and could also endanger a sport fishing industry that is a multimillion dollar yearly segment of the regional economy.

Finally, as pioneering acid rain researcher Gene Likens of Cornell University once pointed out, the damage inflicted by acid precipitation is permanent and irreversible. "No natural means is known," Likens said, "by which an acidified lake might return to its original chemical and biological composition."[17]

With this kind of rap sheet, it's not surprising that acid rain is regarded by many people as one of the most serious environmental threats we have ever faced. A 1981 Canadian Parliament study, for instance, called acid rain "the greatest threat to the North American environment in the recorded history of this continent."[18]

Yet we didn't just wake up one morning and find vinegar splashing on the windowpanes. The problem of acid rain has been a long time in the making. Whoever authored the anonymous seventeenth century English proverb, "To see the rain is better than to be in it," may have known even then that coal smoke was linked to the corroding quality of the rain.

Actually, the problem of acid pollution had its beginnings long before the seventeenth century. British scientist Peter Brimblecombe has noted that as early as 1257, Queen Eleanor of Aquitaine, wife of Henry III, complained of the noxious odors in the British air—odors created by the sulfur compounds that were released by burning coal. By the 1280s, Brimblecombe says, conditions were so bad that two commissions were appointed to do something about the problem. They didn't. For a while, though, Parliament did ban the burning of coal in London during its sessions.[19]

By the 1600s people began to figure out that the polluted air was doing more than simply twinging their noses. Their realization and the economic pressures that dictated their responses offer a fascinating parallel to the acid rain controversy of today.

An energy shortage—in this era, of wood—forced a widespread and rapid increase in the burning of coal in England. Before long writers began blaming coal for damage to health, iron-works, vegetation, stone buildings, and the weather. In 1620 King James I "was moved with compassion for the decayed fabric [of London's Saint Paul's Cathedral]...near approaching ruin by the corroding quality of coal smoke...where unto it had long been subject."[20] In 1661 John Evelyn wrote in his work "Fumifugium" that because of "Clouds arising from those great Fires, the Aer is so distemper'd and such unseasonable and unnatural storms are engendered."[21]

Still, according to Peter Brimblecombe, the English had no real alternative to coal burning. "The continued use of [less polluting] wood was not possible because of the destruction of the English forests....Attempts to limit the use of coal, suggestions for new fuels, and the relocation of industries proved all too futile under the inexorable economic pressures in a society where the powers of industry and capital were growing rapidly."[22]

Thus, faced with a severe coal pollution crisis, the English responded mainly by burning even more coal. In an effort to alleviate the offending coal smoke, they built tall chimneys to release it high above the ground where it could drift away on the wind. That approach, as we shall see later in detail, has striking similarities to our own era's unhappy experience with coal-born acid pollution.

Twentieth-century scientists finally discovered the pollutants in the smoke from the coal fires that were causing all the trouble. Even then, the realization dawned slowly.

The first real clues came in 1911 when Charles Crowther and Arthur Ruston of the University of Leeds in England sampled the rain falling on that industrial city and found that it was pII 3.2, about one hundred fifty times more acidic than the pH 5.7 of normal rain.* U.S. Air Force scientist Helmut Landsburg measured a pH of 4.2 from a single storm in Washington, D.C., and an average pH of 4.0 from eighty-three storms near Boston in 1952 and 1953. Eville Gorham, a Canadian ecologist then studying in England, found acid rainfall

*The pH scale is a convenient way of ranking the relative strengths of various acids. The scale runs from 0 to 14, with 0 being the most acidic and 14 the most alkaline. A substance with a pH of 7, such as pure water, is neutral, neither acidic nor alkaline. The scale is logarithmic, which means that each change of one whole number on the scale corresponds to a tenfold increase or decrease in acid strength. If rainfall in a certain location changes from a pH of 5.6 to 4.6, for example, the rain has become ten times more acidic than it was originally. And if the rainfall changes from pH 5.6 to 3.6, the rain is one hundred times more acidic.

Alkaline

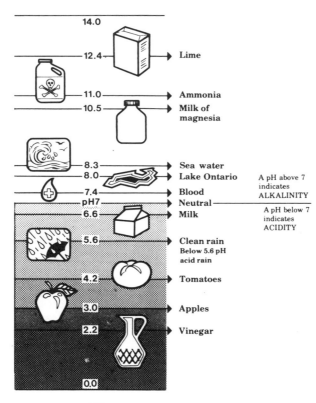

pH	Substance
14.0	
12.4	Lime
11.0	Ammonia
10.5	Milk of magnesia
8.3	Sea water
8.0	Lake Ontario
7.4	Blood
pH7	Neutral
6.6	Milk
5.6	Clean rain
4.2	Tomatoes
3.0	Apples
2.2	Vinegar
0.0	

A pH above 7 indicates ALKALINITY

A pH below 7 indicates ACIDITY

Below 5.6 pH acid rain

Acidic
The pH scale (0-14) measures free hydrogen ions in liquid

Source: Environment Canada

in the isolated English lake country district in the mid-1950s.[23]

But these observations constituted little more than a collection of isolated curiosities until researchers in Sweden and then the United States realized in the late 1960s that the acidic readings were part of a pattern of increasingly acidic precipitation over wide areas of the globe. Svante Oden, a Swedish researcher, analyzed precipitation data collected in Europe since World War II. By 1968 he concluded that postwar industrial expansion on the European mainland had caused a rapid increase in rainfall acidity, not just over Europe but also over Scandinavia, hundreds of kilometers away. Four years later Gene Likens of Cornell, Herbert Bormann of Yale, and Noye Johnson of Dartmouth published a landmark paper documenting a pattern of increasingly widespread and severe acidic precipitation over a large area of the northeastern United States.

Even the work of Likens and his colleagues did not arouse much public attention for several years. For most of the 1970s, discussion of acid rain was almost exclusively confined to the pages of scientific journals where supporters of Likens debated with skeptics and defenders of the electric utility industry.

In 1979, however, a fast-breaking series of events catapulted the acid rain issue onto newspaper front pages, television news programs, and the public consciousness across North America:

● February 1979. The Canadian government reported that as many as 50,000 vulnerable lakes in southeastern Canada are receiving acid precipitation and could be rendered lifeless in less than twenty years. In an interview on National Public Radio's "All Things Considered," John Fraser, then national minister of the environment, called the acid rain phenomenon "the most serious environmental problem that this country has ever faced in my view."

● March 1979. The U.S. Environmental Protection Agency (EPA) reported that acid rain and snow were falling in the pristine Boundary Waters Canoe Area Wilderness of northern Minnesota and that many lakes had already reached the critical threshold at which aquatic life begins to be destroyed. "Say a prayer for the lakes of northern Minnesota," began the article reporting the findings in the *Minneapolis Star*. The EPA study revealed that at least one wilderness lake could be stripped of fish by the mid-1980s and that hundreds more would suffer the same fate during the next several decades if acid continues falling on the region.[24]

● July 1979. The International Joint Commission—an organization charged with monitoring the environmental integrity of the Great Lakes region—heard a series of American and Canadian scientists report that the entire Great Lakes basin was receiving rain and snow that were five to forty times more acidic than naturally occurring precipitation. George Hendrey, a researcher with the

Brookhaven National Laboratory in Long Island, described the widespread damage caused by the acid that is now falling on every watershed draining into the Great Lakes. Many waters that were once thought safe from the effects of acid, Hendrey said, are vulnerable to sudden influxes of acid during snowmelt and to leaching of metals from surrounding soils. Speaking of the fish-killing properties of aluminum set loose by acid rain, Hendrey said: "If they're not dead already, aluminum gives them the coup de grace."[25]

● August 1979. In his environmental message to Congress, President Jimmy Carter called acid rain a global environmental threat and announced a ten-year federal acid rain assessment program.

● October 1979. William M. Lewis, Jr., and Michael C. Grant, two researchers from the University of Colorado, revealed in testimony before a Congressional subcommittee that acid rain was unexpectedly falling and increasing over an area of the Colorado Rockies northwest of Denver.

● November 1979. More than eight hundred people, including representatives from environmental organizations, scientists, government officials, and political leaders, met for two days in Toronto to discuss acid rain. The conference, called the Action Seminar on Acid Precipitation (ASAP), drafted a resolution urging the governments of the United States and Canada to adopt a ten-year program to reduce atmospheric levels of acid-forming pollutants by more than 50 percent.

The Toronto conference was a peak in the movement to publicize and sensationalize acid rain as an issue. In the midst of concern about the dangers of acid precipitation, conference participants were euphoric at the prospect of an issue to revive the lagging environmental movement. The Canadian conference led to another well-attended meeting a few months later in Minneapolis which resulted in a deluge of press and television coverage of the issue. Many Americans and Canadians were made aware of acid rain for the first time.

Even during the heady days in Toronto, however, some environmentalists were all too painfully aware of how ephemeral the moment could be. Arch-conservationist David Brower of Friends of the Earth warned that "the president and almost all of his men rush to worsen [the danger of acid rain] faster, and still worse, to waive the laws that get in the way of the worsening. Down the acid rain drain would go the finest hours of the conservation movement in the course of an entire century. It is as if some higher authority had decreed that the advance of civilization was henceforth to be measured in kilowatt-hours and British Thermal Units."[26]

And Robert Rauch, staff attorney for the Environmental Defense Fund, a nonprofit national conservation organization, cautioned the Toronto delegates not to be deluded by their own unanimity. He said that all the time tested arguments—more research is needed, environmental protection will wreak economic havoc, energy is more important than cleanliness—would be brought to bear again. Rauch didn't have long to wait.

For if 1979 was the year public awareness of the acid rain issue spread throughout the United States and Canada, 1980 was the year the opposition began gaining in strength. In 1980 those who disputed the severity, scope, and even the existence of an acid rain crisis took their campaign from the pages of scientific journals to the media, to the public, and to the U.S. Congress.

The thrust of the argument put forth by a phalanx of electric utilities, coal companies, and their supporters was this: We're not convinced that there is a serious acid rain problem. Moreover, we are not convinced that the burning of coal is the principal cause of whatever problem does exist. And until we can be shown that a batch of pollution from a particular power plant causes a specific form of damage in some other location, then we oppose any effort to impose further expensive regulations on the burning of coal at a time when the nation looks toward that fuel to help meet its energy needs.

The *Wall Street Journal* editorial page was one place the "What Acid Rain?" forces found support. In June 1980, the *Journal*, borrowing from the Electric Power Research Institute—research arm of the utility industry—and the National Coal Association for its information, suggested that "it is difficult to find a direct link between increased coal burning and higher rainfall acidity," and "as to the effects of acid rain on the environment there is also insufficient information."[27]

In a Pittsburgh press conference that same month, William N. Poundstone of Consolidation Coal, the nation's second largest coal company, also emphasized the uncertainty issue. "We looked carefully through the literature and we have not been able to find documented evidence of damage to trees or crops outside the laboratory. There is a serious question of whether the rain has become acidic in the last ten to twenty years. We shouldn't ask the power plants to reduce emissions of sulfur dioxide unless it can be shown to be a cause of acid rain."[28]

Citing the uncertainty of damage as an argument against a move for environmental protection is nothing new in controversies involving resources and those who benefit economically from them. At the core of nearly every environmental battle of the 1960s and 1970s was the debate between those who said the potential for harm was enough to warrant remedial action and those who said harm had to be demonstrated first.

Perhaps the classic case of this kind was the Reserve Mining controversy in which an iron mining company spent more than a decade in court fighting attempts to halt it from dumping 60,000 tons of taconite mining wastes into Lake Superior each day. The company argued that the potential for harm to those who drank lake water containing asbestoslike particles that sifted out of its wastes was not great enough for it to stop the dumping.

The federal courts, however, disagreed. In a decision that reflected hundreds of other cases like it across the country over more than a decade, the court ruled that the damage did not have to occur before the company was required to halt the dumping of wastes. Reserve Mining was ordered to stop dumping and did so finally in 1980.

Because the acid rain issue poses the same question today, it may become a watershed in the history of environmentalism. Abundant evidence exists, as we shall see, that there is a clear potential for ecological disaster resulting from acid precipitation. Yet there is considerable uncertainty about the degree and extent of many of its potential effects. The coal-burning industries, which are the source of most of the compounds that end up in acid rain and snow, have taken the position that proof of actual damage is necessary before any restrictions on coal burning are imposed by the U.S. government.

Thus after more than a decade in which the environment won the benefit of the doubt, the acid rain issue has raised the question yet another time. Within the past few years, however, an entirely new political, economic, and energy climate has evolved. The United States has grown politically conservative as a nation, it is in dire economic trouble, and it is hanging at the end of the most tenuous of energy lifelines. Each one of those developments weighs heavily against the case of the environment.

Late in the summer of 1980 came the first evidence that the tide had turned against the environmentalist philosophy. Robert Byrd, a prominent Democrat who was then the U.S. Senate majority leader, embraced the coal industry's argument on acid rain as the Senate crushed efforts to impose tighter pollution restrictions on Midwestern power plants converting from oil to coal. "We cannot sacrifice [energy] self-sufficiency for speculative environmental interests,"[29] Byrd said at the time.

The election of Republican conservative Ronald Reagan as U.S. president in the fall of 1980 signaled an even stronger movement toward the coal industry position of business as usual until circumstances force a change. Just a few weeks after the election, the *New York Times* reported that under the Reagan administration, "more sympathetic to business and more skeptical of regulation than [the Carter administration], federal policy is likely to swing away from the environmentalism of the 1970s and firmly encourage the burning of more coal."[30]

The key target in this new movement would be the 1970 U.S. Clean Air Act, which established national air quality standards. The provisions of the act, however, do not provide an effective tool in combating acid rain because emission standards apply to the concentration of pollutants in the air only in the immediate vicinity of a pollution source such as a power plant. Most areas of the country have met these standards, which were designed to prevent damage to human health from sulfur dioxide and nitrogen oxides. And yet within those areas, there are still substantial emissions of acid-forming pollutants. In fact, in the Ohio River Valley, which forms the core of sulfur and nitrogen oxide pollution in the United States, power plants are burning high sulfur coal with no pollution controls on their stacks.

The use of tall smokestacks, which is allowed under the Clean Air Act, guarantees that most of a power plant's pollution will be carried off by winds into areas not monitored by local pollution law enforcers. Scientists believe the tall stacks have played a key role in the spread of acid rain in the last two decades. Unfortunately, the use of tall smokestacks shows every sign of remaining the order of the day.

The result, then, of the law's focus on local pollution levels and its tolerance of tall stacks that spread pollution over wide distances has been to create a situation in which many sources of pollution once thought to be under control cause substantial acid damage in distant locations. Strengthening of the Clean Air Act to reduce the use of tall stacks and tightening local pollution standards might have helped alleviate the loading of the atmosphere with acid-forming pollutants.

The advent of Reagan made that out of the question.

The only other remaining hope, according to Gregory Wetstone of the Environmental Law Institute in Washington, D.C., would be for the government to adopt an entirely new approach to regulating air pollution. Instead of defining acceptable pollution levels in a given locale, says Wetstone, the government should require across-the-board "best available technology" pollution controls on all existing sources of sulfur dioxide and nitrogen oxide pollution. For it is these "existing" power plants and factories, many of them built before the institution of clean air laws, that emit the bulk of acid-forming pollutants.

Given the present economic and political realities, government officials are unlikely to try this new approach. "The U.S. government is reluctant to agree to impose pollution control requirements which might make coal a less economically attractive energy source," Wetstone wrote in 1980. "Even in the best of times, efforts to tighten pollution control requirements would face tough political opposition. In today's political climate, competing energy and economic considerations may well prove to be overwhelming."[31]

2.

A Loss of Heritage

They used to walk this same mountain trail forty years ago—five miles in and a thousand feet up to Lake Colden. On summer afternoons the trout fishermen would climb the long path through maple and aspen to the granite hollow below two Adirondack peaks, to the places along the lakeshore where they kept their boats. And when evening rise broke the water into a hundred rippling rings, they would row out slowly and cast their lines toward the hungry trout.

"Fish were breaking everywhere, but I didn't stop to cast until I saw the size of some of them that were cruising on the surface," wrote Vincent Engels. "The air was full of the little smoky wing flies with the kind of brown body and the trout were zig-zagging after them, left and right, tails and backfins and sometimes their snouts too out of the water. I made exactly four casts out there...and caught four fish.

"As these fish were nourished only on insects and crustaceans and smaller trout...they were beautifully colored, deep-bodied, and delicate in flavor—the finest trout I have eaten anywhere."[1]

Engels, author of *Adirondack Fishing in the 1930's, A Lost Paradise*, considered Lake Colden to be one of the best fishing spots in the eastern United States.

Today, Lake Colden is virtually dead.

It has no fish, no flies, no minnows, and no crustaceans; about the only living things left in Lake Colden are algae, sphagnum moss, and one or two kinds of plants. They are the only life forms that can tolerate the acid that floods the lake after rainstorms and with each spring thaw. Lake Colden—along with many other Adirondack lakes—is a casualty of acid rain and snow.

Adirondack Park is a 7-million-acre landscape of wilderness peaks, forests, streams, and lakes in north central New York state. When logging and other development began degrading the mountain region in the late 1800s, the people of New York amended their constitution to designate the Adirondacks "forever wild" and a sanctuary to be protected vigilantly. Wise as they were a century ago, the New Yorkers could not foresee that their park would be ravaged by pollutants borne on the wind, beyond their control.

In the 1930s almost all high Adirondack lakes— those over two thousand feet in elevation— contained fish and were popular fishing spots. About 1950, however, visitors and park officials began to see fewer and fewer fish. A number of causes were suggested, such as excessive beaver activity or windstorm damage to the lake watersheds.

There were attempts to counteract the losses. Helicopters carried young fish to restock the high lakes, including Lake Colden. When the fingerlings were dumped in the center of the lake, they fled almost immediately to the mouths of inlet streams. Several months later they were found dead, and the restocking efforts were abandoned.

Finally, in 1975, a Cornell University scientist discovered why the fish were disappearing. Carl Schofield's survey of the Adirondack lakes revealed that more than half were devoid of fish and most other forms of aquatic life. The explanation, according to Schofield, was acid rain and snow.

The discovery that the Adirondacks were suffering the same fate as areas of Sweden and Norway has been followed by distressing findings in other parts of the world: Ontario, the Boundary Waters region of northeastern Minnesota, the Great Lakes, New England, areas of Nova Scotia, the Rocky Mountains, the Sierra Nevadas, the Appalachians, Florida, and even Venezuela, far to the south of the industrialized countries of North America and Europe, have been added to the vast areas of the globe now subject to the ravages of acid rain.

To understand why some of these areas are vulnerable to the effects of acid precipitation and others are not, it is necessary to know something about acids and how they work. In general, acids are chemical compounds that break apart into smaller pieces. These pieces, in turn, react with and break apart other compounds such as metals and a group of compounds known as bases.

Perhaps the most familiar acid is vinegar, which is a dilute solution of acetic acid. The bubbly reaction of vinegar with the chemical base sodium bicarbonate (baking soda) has been a classic feature of kitchen chemistry labs for years. But the reaction also illustrates a crucial element in the acid rain story, the principle of buffering.

When vinegar comes in contact with baking soda, the two compounds are broken apart and recombined in forms which can no longer react. If, for instance, a capful of vinegar were poured into a bowl of baking soda, the vinegar molecules would react with an equal number of soda molecules, and all of the acid would be neutralized. There would still be baking soda left over, and it is this reservoir of unused acid-neutralizing ability that is known as a buffer.

The acids in acid rain and snow are mainly dilute forms of sulfuric and nitric acid, both of which are much more strongly reactive than vinegar. How these acids affect water, soil, animals, and plants depends on how much buffer is available to neutralize the acids when they fall.

A major source of buffering material is the rock beneath the soil. Limestone and other sedimentary rocks are types of rock rich in chemical compounds that neutralize acids. If a lake, for example, rests on rock with a plentiful supply of these natural buffers, then chances are good that acid precipitation will be neutralized by the dissolved buffers circulating in the lake water.

The problem arises, however, when large areas rest on rock that contains little or no buffer. More than 1.5 million square miles of North America are believed to fall into this category. The bulk of this land rests atop the Canadian shield—a layer of bedrock formed 600 million to 4 billion years ago that stretches from the Arctic Circle across most of Greenland and the eastern half of Canada into the United States just south of the Great Lakes. Other large pockets of buffer-deficient geology in North America include the northeastern and southeastern United States, the Appalachians, the Rocky Mountains, and the Sierra Nevadas—the heart of the most sublime and most valuable country on the continent.

When acid rain and snow falls on these areas, no reservoir of acid-neutralizing compounds is there to prevent acidification; the acid accumulates and becomes more concentrated in the environment. A 1981 report by the General Accounting Office, the investigative arm of Congress, concluded that scientists generally agree that acid precipitation is making some lakes and streams in the United States—mainly in the Midwest and Northeast—increasingly acidic and that the rising acidity has caused some damage to aquatic life.[2]

The mechanism and scope of this destruction is probably the most clearly understood aspect of the acid rain problem. Simply put, the intricate web of aquatic creatures that has evolved and thrived under fairly constant environmental conditions for thousands of years cannot tolerate the massive disruption of acid precipitation. When they can, fish and other creatures subjected to acid insults move to a better neighborhood. Species trapped in acidified areas, however, do not have enough time to evolve and adapt to the new conditions.

As early as the turn of the century, Scandinavian fishermen began to see declines in salmon populations in the highly productive river fisheries of southern Sweden and Norway. In the 1920s, some high mountain lakes in Sweden were found to be highly acidic and devoid of fish. By the 1960s, the commercial salmon fishing industry in southern Scandinavia had been decimated. The director of a Norwegian study of the effects of acid rain on fisheries concluded that "the majority of inland waters...have completely lost their fish populations."[3]

When the connection was made between the dying fish and the increasingly acid rain and snow in Scandinavia, research began into how the acid claimed its victims. Adult fish subjected to acidic environments in the laboratory suffered retarded growth, deformities, and even death. Scientists soon discovered, though, that acid rain threatened the young fish even more seriously than the adults. Carl Schofield of Cornell University says the clue to how acid rain destroys a fish population is that only very old fish are found in lakes that are becoming more and more acidic. "In a normal lake you would expect to find predominantly younger fish," says Schofield. "What happens in an acidified lake is that the age structure shifts toward older, larger, nonreproducing fish. A senile population."[4]

The young, spawning fish and its eggs are most vulnerable to the effects of acid. Increasing acidity disrupts the mechanism by which fish maintain the correct balance of chemicals in their blood and tissues. Specifically, acid interferes with the ability of females to retain the calcium needed to produce eggs. And even if eggs are produced and fertilized, the stress of the acid destroys them.

The killing is insidious and not readily evident until most of the fish are gone. One by one, the "year classes" of fish in a lake or stream disappear. First, there are no one-year-olds; then, no one-year-olds or two-year-olds. Before long the only fish left are those that were born before the acid concentration reached the deadly threshold at which the eggs cannot survive. When they disappear, all the fish in the lake are gone.

Species of fish differ in their ability to resist the effects of acid. As a lake becomes increasingly acidic, however, survival is temporary for most. At pH 6.5, fewer brook, rainbow, and brown trout eggs hatch, and the young grow more slowly. At pH

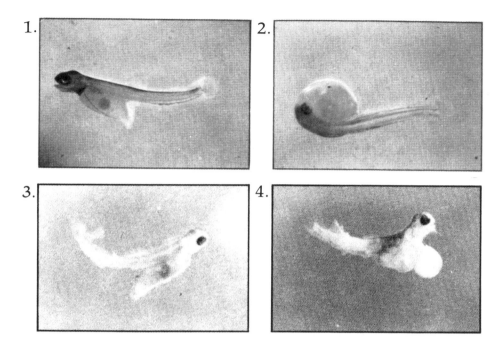

1. A normal, live Atlantic salmon alevin showing body pigmentation and noticeable heart and vitelline vein. 2. During the hatch process this alevin has freed only it's tail, a common condition in waters of pH 4.0 - 4.5. 3. This alevin died in acid waters of pH 4.0. 4. This alevin died in acid waters of less than pH 3.5.

5.5, largemouth and smallmouth bass, walleyes, and rainbow trout are eliminated, and other trout and salmon are in substantial decline. Once a lake slips below an acid threshold of about pH 5.0, even the hardiest fish begin to succumb.

Schofield and other researchers have discovered that acid precipitation poses a double-edged threat for water creatures. Some damage, such as the loss of year classes of fish, is chronic; it is caused by the accumulation of many small doses of acid. Other forms of damage, however, are acute; they are caused by the influx of high concentrations of acid.

In cold climate areas receiving acid precipitation, the winter snows hold a cruel surprise in store for spring. Each winter snowstorm brings to earth a load of acid. As the snow piles up through the winter, the acid accumulates in the snowpack over a lake, river, or stream. When the spring thaw arrives, researchers have found, the first water that melts out of the snowpack is loaded with acid in such high concentrations that it can, by itself, kill adult fish by drastically altering their blood chemistry. In 1975, for instance, thousands of trout died in the Tovdal River in southern Norway after acidic snowmelt caused the pH to plunge into the highly acidic range.[5] Carl Schofield has documented the same phenomenon in the Adirondacks.[6]

In a cruel coincidence, the "slug" of snowmelt acid also hits many species of fish at precisely the

Spring pH Depression of a Stream

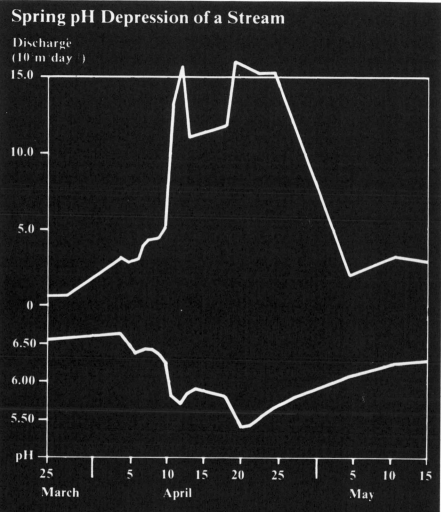

Source: Ontario Ministry of the Environment, 1980

This graph illustrates the 'spring pH depression' of one of the inflowing streams to Harp Lake, a study lake located in Muskoka. The combination of increased spring runoff and acidic melted snow causes the stream pH to drop, producing severe chemical or 'acid shock' effects on aquatic life.

time they are reproducing in sandy shallows or shoals. The adults may be able to escape by moving to deeper water, but the eggs and emerging larvae are left behind to die. A generation of fish dies each spring until there are no fish left capable of reproducing, and the species in that lake, river, or stream disappears.

Near Kenora, Ontario, Canadian researchers have been conducting novel experiments to find out how lakes are affected by acid precipitation. Instead of waiting a decade for a lake to become acidic, they have accelerated the process. At a lake known only as Lake 223, they have added three metric tons of sulfuric acid over the last four years—about the same amount of acid as the lake would have received if it had been located near the metal-smelting city of Sudbury, Ontario. The pH of the lake has dropped from 6.5 to 5.6, which means that the lake has become ten times more acidic than its previous condition. Although that level usually is not considered critical, the researchers have already discovered damage. Fresh-water shrimp and fathead minnows have disappeared, and the numbers of other fish, such as slimy sculpin, have declined sharply. Lake trout embryos have shown an unusually large number of malformations. Finally, toxic metals such as aluminum and zinc have been leached out of the lake bottom rock and are now present in increasing concentrations.[7]

The discovery of increased levels of metals in

acidified lakes explains why fish are being killed in lakes with acid levels that should not be lethal. It also means that many more lakes, rivers, and streams may be endangered by acid precipitation than researchers had thought possible.

Scientists in Canada, the United States, and Europe have measured increases in a number of poisonous metals—aluminum, manganese, zinc, copper, cadmium, mercury, lead, and nickel—in acidified lakes. Research has shown that increasing acidity increases not only the amount of these metals present, but also increases the probability of their methylation, or conversion to organic chemical form. These transformed metals, especially aluminum, are generally more toxic in their converted states.

Carl Schofield has discovered why such mobilization of metals could make lakes in areas with a seemingly adequate supply of buffering chemicals vulnerable to the effects of acid rain and snow. He found that when sulfuric and nitric acid react with soils and release aluminum, this toxic metal washes into lakes. The aluminum-tainted water moves in a layer across the tops of lakes. If fish take it in through their gills, their tissues are damaged, and their body chemistry is disrupted. In effect, they die of asphyxiation. In deeper lakes with a plentiful supply of oxygen, fish can escape the lethal aluminum. In shallower lakes, however, their fate is sealed. "They have the choice of dying from lack of oxygen at the bottom or from acids and aluminum at the top,"[8] Schofield says.

Until the fish in such lakes die, they accumulate and concentrate other liberated metals, such as cadmium and mercury, in their tissues, some to a point that exceeds human health standards. This buildup of metals could be especially dangerous to people living near these lakes who eat many game fish; for example, a number of central Canadian Indian tribes.

The ability of acid to free toxic metals may pose another threat to human health. Since the Great Lakes basin is now receiving from five to forty times the normal amount of acid in precipitation, Canadian and American pollution officials are worried that the tens of millions of people who draw their drinking water from the lakes, rivers, and streams in that watershed may be subjected to small but chronic doses of toxic metals leached out of lake and river sediments and out of water pipes by acid. In some small cities in New York and Vermont, water supplies have been found to contain acid-liberated lead, and increased acidity levels have been measured in two large bays of Lake Huron, a heavily tapped source of drinking water. Municipal drinking water treatment can correct acidity problems before water gets to piping systems, but it does not remove the liberated metals which are already in the water that is tapped by the treatment plant.

The destruction in the aquatic environment caused by acid does not end with fish. Each downpour of acid punches holes in the finely strung web of the aquatic ecosystem. Insects, invertebrates, snails, freshwater clams, plankton, mayflies, and stoneflies are killed or driven away. The variety of plants living in the water is drastically reduced until only a few acid-tolerant species are left.

At three study lakes in the Adirondacks, George Hendrey, a biologist at the Brookhaven National Laboratory, is investigating the effect acid has on the decomposition of dead plants and organic material that falls or is washed into a lake. Hendrey has discovered that acid slows down the normal recycling process and allows debris to accumulate on the lake bottom. In addition, he has found that decomposition changes from the normal aerobic, or oxygen-supported, process to an anaerobic process without free oxygen. The result is obnoxious odor. "Marcy Dam [an Adirondack lake] is one of our most popular recreation areas," says Hendrey. "But the decomposition process has been interfered with to the point where the region smells bad. It stinks."[9]

More important than the smell, however, is the discovery that the slowdown in recycling of dead material, combined with the blossoming of dense mats of mosses, fungi, certain algae, and a few acid-tolerant plants like bladderwort, is causing the lakes themselves to disappear. Lake Colden, the Adirondack fishing paradise of the 1930s, is one of those vanishing lakes. "Filling in is a natural process, but one that occurs on a geological time scale," says Hendrey. "The lakes in the Adirondacks are ten thousand to twelve thousand years old and normally fill in very slowly since most debris is decomposed. In these [acidified] lakes, it's happening extremely fast, not in a geological but in a human time span."[10]

Even creatures that don't spend all of their time in the water appear to be victims of acid rain. The spotted salamander, for example, a species prevalent in eastern North America, emerges from hibernation on early spring nights each year to lay its eggs in temporary pools of rainwater or snow-melt. For its eggs and new embryos to survive, the acidity of that pool water must be within a certain range of tolerance.

F. Harvey Plough of Cornell University has found that as the rain and snow in the north-eastern United States have become more acidic, so have the breeding pools of the spotted salamander. Eggs laid in acid pools are either killed outright or produce deformed embryos that do not survive. Plough calls the future of these salamanders bleak, and he warns that more is at stake than the extinction of one species.

"The significance of widespread failure of salamander reproduction will extend beyond the salamanders themselves,"[11] Ploiugh says. Sala-manders are important predators of mosquitoes and midges, and the absence of salamanders could result in the proliferation of those pests. In addition, salamanders are an important food source for birds, shrews, foxes, and other small mammals. If the spotted salamander disappears, those animals will also be affected, causing a chain reaction that reaches ever deeper into the ecosystem.

In addition to its effects on the animals and plants of aquatic ecosystems, acid precipitation may affect the viability of other forms of plant life. Although acid rain has not been proven to have a significant effect on plant growth in forests, on the range, or on the farm, research indicates that where acid is plentiful, plant growth is not.

Acid could harm plants in a variety of ways:

● Destruction of leaves and other plant surfaces. F. Herbert Bormann of Yale University, one of the foremost acid rain researchers in the United States, found that acid rain could harm birch tree leaves so severely that some trees were left without enough leaf surface to sustain normal growth.

● Interference with normal reproductive processes. Some scientists believe that studies showing a stunting of seedling growth by acid suggests that, like fish, plants are especially susceptible to the effects of acid when they are young and going through a series of critical phases of growth.

● Synergistic, or reinforcing, interaction with other pollutants, such as the combined effects of sulfur dioxide, ozone, fluoride, and soot.

● Increased susceptibility to drought and other environmental stresses. Damage to plant cells could disrupt the process by which plants retain essential moisture.

● Increased susceptibility to disease. Laboratory and field studies have indicated that acid rain can make forests vulnerable to a variety of plant diseases. David Johnson, biologist for the Adirondack Park Agency, has observed that in the last decade, sections of the forest have begun dying from diseases that had never been seen before in the wild. "We're seeing forest diseases now that for years and years were thought of as being strictly nursery diseases," says Johnson. "Now serious pathogens are showing up in the wild and we've had wipeouts of trees."[12]

● Disruption of the delicate soil environment that supports plant growth. The same mechanism that washes poisonous mercury, lead, cadmium, and other metals from the soil and into lakes also removes the nutrients plants need. In normal soil, nutrients are collected and held as they are received from weathered rock, rain, or the atmosphere. They are taken up by plants and returned later to the soil as dead matter in a continuous recycling process. Various steps in that process are made possible by bacteria that either break down organic material or convert nutrients into forms plants can use. Acid precipitation rips apart that carefully arranged system because many bacteria and other microorganisms cannot function or even survive in strongly acidic environments. In addition, acids prevent soil particles from retaining nutrients long enough for plants to use them.

● Interference with photosynthesis, the crucial growth process in plants. Potentially, this is the most serious harm that acid could inflict on plants and, indirectly, to humans. Photosynthesis is the process by which plants convert the light energy of the sun into chemical energy, in the form of simple sugars, for their growth. The key to photosynthesis is the chemical chlorophyll, which absorbs sunlight and enables plants to produce oxygen and hydrogen. Some studies have shown that acid alters chlorophyll so that it cannot function. Other studies have shown that when acid interferes with photosynthesis, seeds, fruits, roots, and tubers are the parts of the plant that suffer most. These are also the parts that humans use most for food.

These potential effects of acid on plant growth may have a severe impact on important U.S. cash crops. George Hendrey and other researchers have found that soybeans—a cash crop worth $13.6 billion to U.S. farmers in 1980—could be damaged by acid rain's interference with necessary soil microorganisms. They discovered that acid prevented the normal formation of the bacteria nodules on soybean roots which take nitrogen out of the air and make this essential nutrient available to the plant in the soil.[13]

A 1979-1980 Department of Energy study showed that soybean yields were cut drastically by exposure to concentrations of sulfur dioxide—the

main chemical precursor of acid precipitation—
even within the legal limits set by the federal
government. Seed size and weight and the number
of pods on each plant were cut by as much as 45
percent.[14]

Some scientists believe that acid rain may
prove beneficial for some crops. In an experiment
done at the Environmental Protection Agency
laboratory in Corvallis, Oregon, tomatoes, straw-
berries, and corn were grown in artificial sulfuric
acid rain of concentrations as high as pH 3.0—an
acid strength between lemon juice and vinegar.
These crops grew well, while other crops, such as
beets, carrots, and radishes, grew poorly. Still
others grew adequately but suffered leaf damage
that would make them unsightly and therefore
unmarketable. The researchers concluded that
acid rain would most likely harm leafy vegetables
and vegetables whose seedlings were monocoty-
ledons as opposed to the two-leaf seedlings of such
crops as corn and tomatoes.[15]

The multi-billion dollar a year timber, pulp,
and paper industry in North America and Scandinavia may also be affected by acid precipitation. In
the northeastern United States, scientists suspect
that acid damage may be responsible for a recent
decline in forest productivity. Researchers at the
University of Vermont and Yale University have
investigated a major die-off of spruce trees during
the last fifteen years in an area of Vermont's Green
Mountains that is subjected to frequent and

intense episodes of acid precipitation and acid haze. The researchers have narrowed the cause of the tree deaths to some form of environmental stress but have not been able to implicate acid directly. They note that acid does fall on the area and that heavy metals such as lead, copper, and zinc are accumulating in the forest soils. "Since the environment of high elevations is normally fragile," they conclude in a research paper, "it is possible that recent atmospheric pollution is sufficient to tip the balance of trees already growing in a stressed situation."[16]

Although scientists are still trying to ascertain and detail the damage acid causes in the plant world, the damage it is causing in the human realm is better known. To understand how acid affects human health, it is necessary to examine the two forms in which acid falls, dry and wet. Wet acid in rain and snow has one set of deleterious effects. Dry acid particles in the air we breathe, which may account for half of all the acid in the environment, has a different set. Thomas Seliga, an Ohio State University atmospheric scientist, says that the dry and wet forms of acid "are all related and all very much a unity or whole, and it all has to be dealt with in that fashion."[17]

The urgency in his statement comes from studies such as the one conducted in 1981 by one hundred researchers from eight U.S. universities. They examined the Ohio Valley, an area with many coal-fired power plants that

produce about one-fourth of the acid-forming
pollution in the United States. Not surprisingly,
the Ohio Valley also has the highest levels of
airborne acid sulfate particles in the nation. These
particles can be breathed into the lungs where they
actually burn lung tissue and are believed to
aggravate heart and lung diseases. The $4.3 million
study concluded that if the level of coal-fired
electrical generation in the region continued un-
changed, 163,000 people would die within the next
twenty years because of their exposure to the acid.
Though the study's conclusions were called grossly
erroneous and unsupportable by the electric power
industry,[18] another recent study by the Brookhaven National Laboratory found that airborne
sulfates may be responsible for 5 to 8 percent of all
deaths in some parts of the United States.[19]

In addition to the hazards posed by acid to
human health, acid damage to materials in the
United States each year is estimated to total more
than $5 billion.[20] Of the materials corroded by
acid, metal alloys and stone are the most economically significant; in fact, steel, nickel-plated steel,
antirust coated steel, and even galvanized steel, are
corroded by acid rain. British railroad officials
attribute one-third of the replacement cost of steel
rails to acid corrosion.[21] Paint manufacturers
acknowledge that acid rain is breaking down
paints, particularly oil-based and automobile
finishes. Some scientists now estimate that as
much as half of the rust in American cars is caused

by acid rain and slush, while the other half is caused by road salt.[22] Aluminum, which is widely used in siding, rain gutters, storm window frames, and lawn chairs, is especially susceptible to acid rains of the strength now falling over much of the Northern Hemisphere. Even the nylon in umbrellas is broken down by acid rain.

The effects of acid on building stones, however, are among the most visible and the most dramatic. In the geologic time frame, the naturally occurring carbonic acid of rain unpolluted by human activity was the agent of decay of stones in the earth and, later, of stones in buildings. It worked very slowly, wearing the surface down a grain at a time in a reaction that could occur only when the stones were wet.

Today sulfur dioxide and its acidic by-product are the dominant agents of decay. They work quickly to transform the sturdy surfaces of marble, limestone, and other building stones into a crust of gypsum plaster that crumbles to the touch, at the vibration of traffic, or with the wash of the next rain. Worse, windborne particles of acid or sulfur dioxide can decay stone surfaces that are not directly exposed to the rain because such particles need only water vapor to trigger their corrosive action.

Around the world, centers of history and tradition are being steadily eaten away: the Taj Mahal of India, the Colosseum in Rome, and the Lincoln Memorial in Washington, D.C. "We can

even see streaking on the Washington Monument
—possibly the result of acid rain's grooving and
pitting effects. Rock that might last hundreds of
years is lasting only decades,"[23] said Kenneth J.
Hood, a former official of the Council on Environmental Quality and the executive secretary of
the inter-governmental Acid Rain Coordination
Committee.

The city of Venice, Italy, employs people who
do nothing but locate and remove the exterior
decorations of buildings that have been struc-
turally weakened by acid. Statues in Germany
carved in 1702 showed little deterioration in 1908;
the same works today are badly eroded. In Rome
photographs of the Column of Marcus Aurelius on
the Piazza Colonna reveal that many of the
sculpted figures in its intricate relief that were
intact in 1950 have been mutilated and become
unrecognizable. "The same is true of the Column
of Trajan, the Arch of Constantine, the Arch of
Septimus Severus, the Arch of Titus, the Temple
of Romulus and the other monuments standing
between the Piazza Venezia and the Colosseum,"
reported the *New York Times* in 1980. "At the
present rate of decay virtually all the sculpture on
these monuments will disappear within twenty
years."[24]

Edward McManus, an architectural conserva-
tor with the National Park Service, conducted an
inspection of the Statue of Liberty in 1980 to
determine how much damage two protestors had

inflicted while climbing the structure. He found that the climbers had done little damage compared to the effects of acid rain. According to McManus, "The environment has become much more hostile" in the forty years since the statue had been last overhauled. "There's a tremendous problem with acid rain that may have caused the material to deteriorate at a faster rate than we thought,"[25] McManus said. He was referring to holes in the copper surface of the statue and large patches of flaky corrosion.

With a huge expenditure of money and innumerable hours of skilled labor, some of this damage to human artifacts can be repaired. Most, however, is irreparable by any known technology. And that is perhaps the greatest cost exacted by acid rain—the loss of heritage.

Consider the comments of Dimitrios Nianias, Greece's former Minister of Culture, about the statues that were removed from the Parthenon in Athens—the classic features of eyes, nose, and cheekbones that have dissolved into an acid blur and have been replaced with fiberglass replicas. "Those maidens looking out over Athens had witnessed some of history's most dramatic developments," Nianias said. "Starting from the fifth century B.C. they saw Plato, Aristotle, Alexander the Great, conquering Roman emperors, the Byzantine era and so many wars fought at their feet. They survived all other catastrophes to be defeated by modern day pollution."[26]

3. Out of Sight, Out of Mind

Two mammoth bunkers of concrete and steel rise above the fields of corn and soybeans near Becker, a small town in the central Minnesota farm country. Deep inside, constant fires burn at temperatures that hover around 3000° F. Searing hot exhaust gases roar from the mouth of the 650-foot-high smokestack and rise into the summer sky, boiling and curling until they are swept away on the wind in a pale, yellowish plume.

Acid rain is born here in the fiery heart of the coal-fired boilers of a Northern States Power Company power plant, and in all the other places fossil fuels—coal, oil, or natural gas—are burned. It originates not only in the emissions of power plants, but also in those of the metal ore smelters of Canada, the petroleum refineries of California, the steel mills on Lake Michigan's south shore, the furnaces in homes bound by northern winters, and the internal combustion engines of every one of the hundreds of millions of cars on the world's highways.

In short, acid rain is the unwanted child of an energy-intensive industrialized culture. Since the first stirrings of the Industrial Revolution, the compulsive drive to use energy to replace labor, to create comfort, to provide convenience, and to increase profits has made clean air a memory in many parts of the world. And in that fouled air—a great equalizer that makes the visibility at the top of Vermont's highest mountain frequently as bad as downtown Los Angeles—are the ingredients of today's global acid rain problem.

Scientists studying the acid rain phenomenon believe that the single most important cause of the acid precipitation falling today on North America and western Europe is the burning of coal for electric power. In addition to its basic elements of carbon, hydrogen, and oxygen, coal contains quantities of sulfur and metals such as mercury, cadmium, zinc, and arsenic. In the inferno of the power plant boiler, coal molecules are ripped apart and rearranged to release the energy needed to make the steam that turns the electric generators. As a by-product of this process, atoms of sulfur from the coal and nitrogen from the air are paired up with atoms of oxygen to form sulfur dioxide and nitrogen oxide gases which are released into the air through the plant smokestack. These are the gases that are chemically transformed in the atmosphere to form dilute sulfuric and nitric acids, the major culprits in the acid rain story.

The rain actually was slightly acidic long before

humankind began to use the earth's natural resources to produce energy on a massive scale. When sulfur dioxide and nitrogen oxides from industrial smokestacks are released into the atmosphere, they join an already complex chemical cauldron. In addition to the predominant gases in the air—oxygen, nitrogen, and carbon dioxide—a number of other elements are present, all of which are involved in the chemistry of rain and snow.

Water in the atmosphere comes primarily from

Sources of acid rain

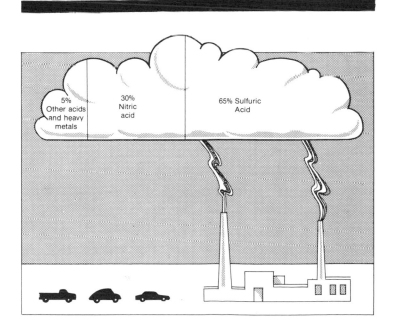

the evaporation of oceans, lakes, rivers, and from transpiration, the process in which plants give off water vapor. The water vapor rising from the earth is pure. Once it has condensed into droplets in the atmosphere, however, it quickly combines with gases and other substances.

Carbon dioxide, most of it exhaled by plants, mixes with moisture in the atmosphere to produce a mild acid called carbonic acid, the same chemical that gives soda pop its fizz. Thus rain in its natural, uncontaminated form is changed into a mild acid before it hits the ground.

During past eras of earth history, this mild acid had an important function in the ecological scheme. The primeval "acid" rain had just enough strength to dissolve rocks in the earth's crust in a slow process that released the minerals and nutrients needed to sustain plant life. In a continuous self-sustaining cycle, plants helped create the acid that released the food that fed the plants.

In an atmosphere unpolluted by human activity, other influences on the chemistry of precipitation would be present; for example, the natural sources of acids stronger than mild carbonic acid. Volcanoes, forest fires, and lightning release huge quantities of compounds that can be converted into sulfuric and nitric acid. In pre-industrial times, however, the amount of these acids was more or less constant, and it was distributed on a regional or global scale so that the acids were neutralized by natural atmospheric processes. Ammonia gas created by decaying organic matter on earth, for instance, forms ammonium ions in the air that counter acidification. Ocean spray, dust, and other particles lifted into the atmosphere by winds also contain elements that neutralize acids.

With the advent of the Industrial Revolution, large-scale burning of fossil fuels began to upset nature's carefully balanced system of rain and snow. As more furnaces were built and stoked, the amount of acid-producing pollutants in the atmosphere multiplied. The emissions of these pollutants increased dramatically after World War II when industrial machinery and power plants began consuming enormous quantities of fossil fuels and polluting the air with their residue.

Today the United States alone is responsible for releasing 56 million metric tons of sulfur dioxide and nitrogen oxides into the atmosphere each year. Europe, not including the Soviet Union, adds nearly 60 million metric tons.[1] Worldwide, scientists estimate that human activity is responsible for more than one-third of the acid-producing sulfur compounds[2] and from 10 to 15 percent of the acid-producing nitrogen compounds.[3]

If the air masses over land, particularly in the more heavily populated and industrialized Northern Hemisphere, are examined separately, then the magnitude of humankind's contribution to these pollutants is 50 percent or more.[4] And if the scope is narrowed even further, scientists estimate that about 90 percent of the acid-producing sulfur compounds in the air over eastern North America arises from human activity.[5] One industrial complex, the huge Inco Ltd. copper and nickel smelting plant in Sudbury, Ontario, has produced as much sulfur dioxide in the last decade as the total produced by volcanoes during billions of years of earth history.[6]

Such a formidable concentration of human-made air pollution may have grave consequences. "The atmosphere still possesses a truly formidable capacity to dilute, disperse, and destroy an enormous list of substances man chooses to

Increase in sulfur dioxide and nitrogen oxide pollution

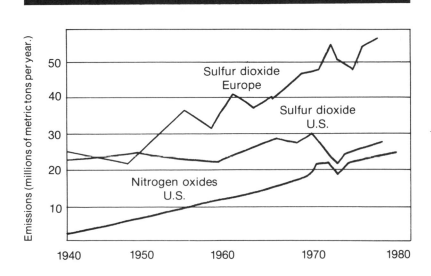

indifferently discharge into it," Earl Dunbar
Murphy, a Ohio State University professor, told
an international symposium on acid rain in 1976.
"The renewing environment's powers have till
now indulged man in his use of them as sinks, but
the time of that indulgence is drawing rapidly to a
close, if it has not done so already."[7]

In recent years the atmosphere's capacity to
absorb pollutants has indeed been pushed too far.
Deep in the ice of the Arctic, samples of precipitation that fell about two hundred years ago have
been measured with a pH range of from 6.0 to 7.0.[8]
In 1979 scientists analyzed samples of new fallen
snow at Point Barrow, Alaska, well above the
Arctic Circle, and found a pH of 4.7.[9] Thus in two
centuries the rain and snow falling in one of the
most remote regions on earth had become one
hundred times more acidic. The researchers have
concluded that the source of the acid snow and the
acid haze routinely observed in Arctic air is
industrialized western Europe, about six thousand
miles away across the North Pole.[10]

The advance of acid precipitation in the Arctic
is startling enough, but in other areas of the
Northern Hemisphere it has been overwhelming.
Throughout most of the eastern United States,
southeastern Canada, and western Europe, the pH
of the yearly precipitation is between 4.0 and 4.5—
rainwater so acidic that fish cannot survive in it.[11]
The average pH of New York City rainfall is 4.28,[12]
while the average pH readings of precipitation in

rural areas of New England can be even lower. The most acidic rainfall ever recorded—pH 1.5—occurred in Wheeling, West Virginia, during the infamous 1978 storm.[13]

Just thirty-five years ago, the rains in most of these areas were either barely acidic or slightly alkaline. By the mid-1950s, a map of areas of the United States with a rainfall pH of less than 4.5 included part of the Ohio Valley, central Pennsylvania, western New York, and a portion of New England—the areas in which most of the nation's industrial activity was centered at the time. The most recent maps, however, show that the area of highly acidic rainfall has grown many times in size and now reaches far into the South, the Midwest, and Canada.[14] A map of Scandinavia and northwestern Europe shows an almost identical growth proportionately in the area affected by acid rain.[15]

"There is a regional pattern of acid precipitation covering almost all of eastern North America," wrote acid rain researcher Charles V. Cogbill of Cornell University in 1976. "This [area of acid rain] is spreading to the south and west with some intensification at its center in the northeast United States."[16]

Close to the center that Cogbill mentioned is the Hubbard Brook watershed in New Hampshire. A long term, in-depth study of the geology, chemistry, and ecology of this forested area began in the early 1960s, and it provides the longest known record of the acidity of rain and snow in the

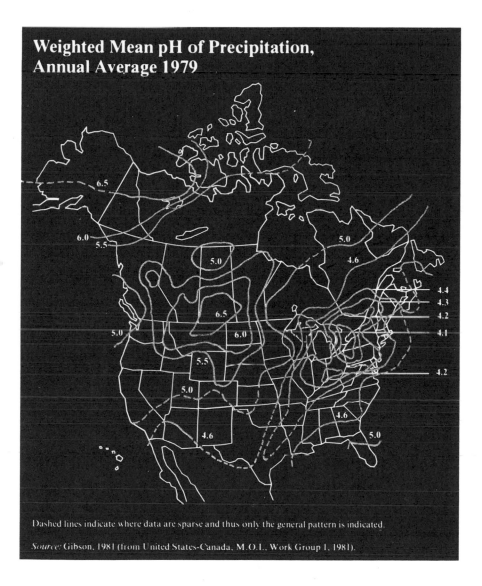

Weighted Mean pH of Precipitation, Annual Average 1979

Dashed lines indicate where data are sparse and thus only the general pattern is indicated.

Source: Gibson, 1981 (from United States-Canada, M.O.I., Work Group 1, 1981).

United States. The Hubbard Brook watershed is a remote area of the White Mountains far from any industry or large city. Since 1964 it has been steadily drenched with precipitation between pH 4.0 and 4.2 and has been subjected to frequent storms of much higher acidity.[17] Gene Likens of Cornell University, F. Herbert Bormann of Yale University, and their colleagues in the Hubbard Brook study have found that while sulfuric acid is the dominant acid in the rain and snow in the area, the amount of nitric acid in the precipitation has greatly increased in the last ten years. This change corresponds to the increased amount and spread of nitrogen oxide emissions from power plants and automobiles during the same period.

In California nitrogen oxide emissions have emerged as a serious problem in the San Francisco Bay area and in the Los Angeles basin. The yearly mean pH of the rain in Pasadena, for instance, has been measured at 3.9, and one storm was recorded at pH 2.8.[18] Scientists have found that the proportions of sulfuric and nitric acid in California precipitation—roughly two-thirds nitric to one-third sulfuric—are the reverse of what is found in the eastern United States. The explanation, according to California Institute of Technology (Cal Tech) scientists who have been studying the acid rain problem on the West Coast, is the extremely high concentration of cars, which emit nitrogen oxides.[19]

The Cal Tech scientists made another important discovery. Researchers suspected that many particles, or acrosols, in a mass of pollution eventually fall to earth under the force of gravity before they are scavenged by rain or snow. In some areas, say the researchers, this dry deposition, or acid smog, may be at least as damaging as acid rain. In a 1979 California study, their theories were confirmed.

The Cal Tech researchers found that about twenty times more acid reaches the ground in particle form than in rainfall because in California rainstorms are usually not frequent.[20] California, then, receives the worst of both worlds: dry acid on trees, cars, plants, and in human lungs all the time; and when it does rain, large quantities of accumulated acid are washed out of the dirty air.

The acid rain situation in Canada is quite different from that in California or the eastern United States. Canada does not have as many people, power plants, or automobiles as the United States, and yet acid rain there has become so severe that Canadian government officials call it the most serious environmental problem facing the nation.

Each year from 7 to 8 million tons of sulfur dioxide fall on a vast area in eastern Canada. Because much of this land rests atop the buffer-deficient Canadian shield, it is poorly suited to withstand the effects of acid precipitation.[21] Environment Canada, the Canadian counterpart of the U.S. Environmental Protection Agency, estimates that at least one hundred forty lakes in Ontario have been stripped of all species of fish by acid and that as many as four thousand more have lost all bass and trout.[22] Farther east in Nova Scotia, the situation is also grim. A 1980 survey by Dr. Walton Watt of the Canadian Department of Fisheries and Oceans reported that nine Nova Scotia rivers can no longer support salmon reproduction and that at current rates of acid deposition, eleven more will be without salmon within twenty years.[23]

Until recently, scientists and officials assumed that since the Canadian areas afflicted with the most acidic rain were in southern Ontario and the eastern provinces—all of which are downwind of industrialized America—the United States must be responsible for most of Canada's acid woes. In 1979, however, a study sponsored by both nations of long range pollution transport indicated that the two countries share equal blame for Canadian acid rain.[24] Pollution from the midwestern United States is indeed carried north across the border, but it mingles with pollutants of the Canadians' own making.

Even though Canada has fewer cars and coal-fired power plants, it does have a gigantic metal smelting industry—an industry that processes ores containing not just copper or nickel but substantial amounts of sulfur as well. As a result, Canada is responsible for more than 5 percent of the industrial sulfur dioxide pollution in the global atmosphere. A full 1 percent of the world's sulfur dioxide pollution comes from the Inco Ltd. smelter in Sudbury, Ontario.[25] Canadian scientists believe that sulfur dioxide pollution from smelters like Inco contributes to acid problems farther east in the country, and they have acted to place limits on acid-forming emissions.

The discovery of acid rain on the United States west coast and in eastern Canada was not surprising considering how close those areas are to major concentrations of people, cars, and industry. But what was unexpected and especially disturbing to scientists was the discovery in 1979 that acid rain was falling in the Rocky Mountains.

William M. Lewis, Jr., and Michael C. Grant of the University of Colorado reported in the journal *Science*: "Our data give evidence of surprisingly low precipitation pH in the Colorado Rockies, suggesting that changes in the precipitation chemistry may be more widespread than presently realized and that the acid rain phenomenon. . .may be typical of significant portions of the western United States."[26]

Lewis and Grant also found during their three-year study that rain and snow were becoming significantly more acidic each year.

If acid was falling on the Rockies, isolated by more than a thousand miles from the highly urbanized and polluted West or East coasts, where was it coming from? To find out, the researchers collected and analyzed rain and snow samples in an area near the Indian Peaks Wilderness Area in Boulder County, about thirty miles northwest of Denver.

The Denver-Boulder-Fort Collins strip along the Front Range of the Rockies is the closest populated and industrialized area to the precipitation sampling station just below the Continental Divide. The next closest centers of pollution in the region are Salt Lake City, 360 miles to the west, and Phoenix, 540 miles to the southeast. Lewis and Grant analyzed detailed studies of weather patterns and found that most air moving over the station came from the northwest and not from the more polluted areas to the east and south. Yet in three years the pH of precipitation falling at the mountain station had decreased from 5.43 to 4.63; in other words, it had become almost ten times more acidic.

The scientists noted that the increasingly acidic trend "cut across seasons and across years with very different weather patterns."[27] They found that nitric acid was apparently responsible for most of the increase in acidity, and yet they were unable to match low pH storms with weather coming from the close-by Denver area, which is heavily polluted with nitrogen oxides from automobile emissions.

Lewis and Grant could come up with only two possible explanations for the acid rain in the Rockies: either there is some complex, never before encountered way that pollutants are transported from the Denver area against the prevailing weather patterns, or else "very widespread changes in precipitation chemistry are presently occurring in the western United States because of significant increases in the release of nitrogen oxides from multiple sources throughout the West."[28]

Most acid rain researchers tend to believe the second theory—acid-forming pollution from the Los Angeles basin is mixing with pollution from other western sources such as power plants and smelters. If they are right, recent and future increases in urbanization and coal-burning in the West may jeopardize some of the most spectacular and valuable wilderness land in the world. At least fifty-eight national wilderness areas and parks are located in areas that are both receiving and are sensitive to the effects of acid rain.

Still, there is the question of why acid rain seems to fall only in some places and not in others. The evidence suggests that the pollution ingredients for acid precipitation are spreading widely across North America. Still, the rain may be many times more acidic than normal in Pennsylvania or Colorado, while it is perfectly fine in Indiana or Iowa or Texas. In the areas unaffected by acid rain, something must be intervening in the rain-forming process before acids in the sky can fall to the ground.

Erhard Winkler, a geologist at the University of Notre Dame, suggested in the mid-1970s that in some places the effects of two ecologically disruptive human activities—industry and agriculture—counteract each other.[29] Winkler theorized that dust from erosion of plowed fields and grazing lands (and to a lesser extent from construction excavation and natural flood plains) is lifted into the atmosphere where it chemically neutralizes acids. The rate of dust deposition on the United States is believed to have at least tripled since farming and construction began in the American plains, according to Winkler. Dust can be lifted upward from bare, plowed, or overgrazed ground to a height of 1500 to 2000 meters by turbulent surface winds, such as those that occur along high pressure fronts. Tilling and earth-moving activities also help boost dust, especially larger particles, into the air. Photographs of North America taken by satellite frequently show large cloudlike areas of dust in an area stretching from Texas north into the southern plains.

Dust can be divided into a few general chemical categories. In terms of acid rain, the most important are the carbonates, such as calcium carbonate from limestone. Winkler believes that once these dust particles are swept into the atmosphere and are carried along over many miles, they sink so slowly that, in the presence of rain or high humidity, they have plenty of time to react with and neutralize acids. In one reaction described by Winkler, calcium carbonate particles react with sulfuric acid in the air to form a neutral compound, calcium sulfate, otherwise known as gypsum.

Winkler said that areas such as New England receive highly acidic rain because densely forested land does not produce nearly as much airborne dust as that produced by the agricultural prairie states. And since dust whipped up in Indiana falls to the ground well before it reaches the skies over New Hampshire, sulfuric and nitric acid particles there are not neutralized.

Recently, other scientists have backed up Winkler's theories about the role of dust in acid rain. University of Minnesota researchers, plotting the pH of rainfall in an arc across the state of Minnesota from the farmlands in the southwest to the wooded lake country in the northeast, found that acidity increased greatly not far beyond the point at which forested land began. The line of demarcation was crucial, they discovered. On the plains side the rain was normal to slightly alkaline; on the forest side it became acidic.[30]

Besides the increased use of fossil fuels to support post-World War II industrial growth, two other key developments dramatically hastened the spread of acid rain in the past thirty years. Ironically, both of these happened because of well-intended efforts to control pollution rather than aggravate it.

The first of these was an attempt to deal with the most obviously offensive element of air pollution: soot. Electronic precipitators, devices designed to electrically filter out particles from a column of smoke, began to be installed in smokestacks in the 1950s. With most of the large chunks taken out of the smoke, the air near factories and power plants looked and smelled cleaner, and window sills did not turn dingy black as quickly as they had in the past.

The sulfur and nitrogen oxides were still pouring out of the stacks, however, and they were still being converted into acids. Not until the 1970s did scientists realize that the advent of precipitators had made the acid problems worse. Soot particles, they discovered, are able to neutralize acid molecules that form in the plume of pollution from a smokestack and turn them into harmless compounds. Since the precipitators removed the acid-scavenging soot, more acid made its way into the air.

Even with the increased acid caused by the soot filters, acids forming in industrial pollution tended to fall to earth relatively close to their sources because smokestacks were not tall enough to discharge the chemicals at a height where the winds could carry them long distances. Then another so-called advance in pollution control technology quickly transformed acid rain from a local to a regional and finally a global problem.

Two often-invoked pollution control adages, "Out of Sight, Out of Mind" and "The Solution to Pollution is Dilution," have given rise to practices such as dumping pollutants into a river and waving goodbye as they headed downstream to become someone else's problem. These approaches

were tempting to those responsible for and suffering from severe air pollution problems in and close
to large cities. Air pollution in England was so bad
in the 1950s, for example, that planes often had a
hard time landing at Heathrow Airport. Some
pollution episodes—especially those involving
high levels of sulfur dioxide—were downright
deadly. The answer seemed obvious: Build the
smokestacks higher so that the smog will travel
farther before it falls to the ground and people
have to breathe it.

And build them higher they did. In 1955 only
two stacks in the United States were higher than
600 feet (slightly taller than the Washington
Monument). But with the impetus provided by the
U.S. Clean Air Act of 1970, polluters turned to tall
stacks. By 1975 at least fifteen stacks higher than
1,000 feet had been built worldwide, several of
them in the heart of the North American industrial
belt.[31] Many others were built between 600 and
880 feet, most of them on coal-fired power plants.
The world's tallest stack, on the Inco smelter at
Sudbury, Ontario, is 1,250 feet high—taller than
the Eiffel Tower and almost as tall as the Empire
State Building.

In 1940, according to the U.S. Environmental
Protection Agency, power plants accounted for
only 3 million tons of the estimated 20 million tons
of sulfur dioxide discharged into the air in the
United States. The rest came mainly from residen-
tial and industrial boilers with short smokestacks.

Today, two-thirds of the more than 30 million tons of sulfur dioxide discharged each year is emitted by power plants with tall stacks.[32] The result is that these pollutants are now capable of penetrating any pressure inversion layer and circulating for long distances with the immense air masses that form our weather systems.

Gus Speth, chairperson of the United States Council on Environmental Quality, talked about high stacks in a speech to a conference on acid rain held in Toronto in 1979:

"One electric utility [American Electric Power] went so far as to take out newspaper and magazine ads back in 1973 bragging that it was a 'pioneer' in the use of tall smokestacks on its power plants to

Chimneys are some of the tallest human-made structures

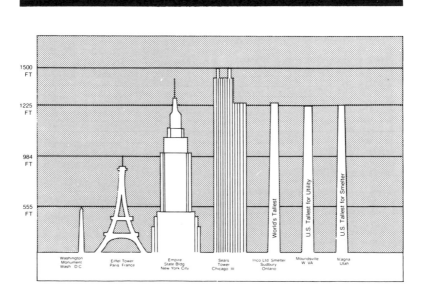

'disperse gaseous emissions widely in the atmosphere so that ground level concentrations would
not be harmful to human health or property.'

"The company claimed that gases from tall
smokestacks 'are dissipated high in the atmosphere, dispersed over a wide area and come down
finally in harmless traces.' It went on to blast
what the company called 'irresponsible environmentalists' who insisted on tough emission standards
at the source of pollution, charging that we were
guilty of 'taking food from the mouths of the
people to give them a better view of the
mountain."[33]

Could the power company be right in asserting
that emissions from tall stacks "come down in
harmless traces?" Since the prevailing winds in
the United States are generally from west to east,
one might assume that the acid-bearing pollution
from power plants and factories in the east central
United States would be blown out to sea. Some is,
but unfortunately a great deal more is not.

Emissions of sulfur dioxide and nitrogen oxides
from tall stacks are captured by the winds of
weather systems. Depending on the strength of
the system and the time of year, the chemicals can
be carried hundreds of miles from their source. In
the summer high pressure systems drift lazily
across the center of North America. The winds in
these systems move in a clockwise direction,
scooping up pollutants from the heavily industrialized Midwest—Cleveland, the Ohio River Valley,

Gary, Chicago, and Saint Louis—and transporting
them west and north through Illinois, Wisconsin,
Minnesota, and across the border into Canada.
Each power plant, each factory, each traffic rush
hour along the way adds its share of chemicals to
the growing mass of pollution that meteorologists
call a "hazy glob."

"Emissions from large coal-fired power plants
in the Ohio Valley, for example, drift lazily along
with this flow, with the sun beginning to 'bake'
those emissions," says meteorologist Walt Lyons.
"What was initially colorless gas is now being
converted by the sun [and also by moisture and
other chemicals] into minute particles, including
small specks of sulfuric acid. That's where the haze
is coming from."[34]

Researchers have been able to watch the
formation and spread of these areas of haze
through satellite photographs and by measuring
changes in visibility at every airport east of the
Mississippi River over a period of days. The globs
form first over the areas where coal-burning
plants are most concentrated and spread like
amoebas to cover most of the eastern United
States.

One group of researchers intensively studied
the contribution of pollution from the Saint Louis
metropolitan area to these "hazy globs." In 1974
and 1975, the scientists mounted instruments in
two airplanes, three piloted balloons, and a van
and traveled along with clouds of pollution that

Source areas of
primary pollutants

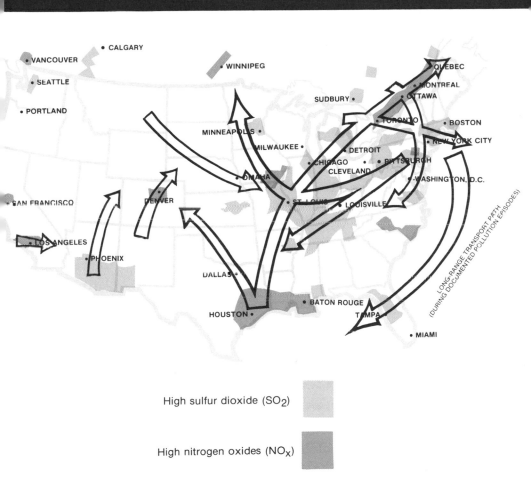

High sulfur dioxide (SO$_2$)

High nitrogen oxides (NO$_x$)

came from individual power plants and from the metro area as a whole.[35]

The mass of pollutants from all sources in Saint Louis—a cloud at times twenty-five miles wide— lost about two-thirds of its sulfur dioxide within sixty miles of the city through filtering and absorption by the ground, trees, buildings, and other objects. Still, a substantial amount of this pollution, like that from the power plants alone, ended up as sulfuric and nitric acids that continued on a lengthy journey.

When the researchers tracked the pollution from individual power plants, they were able to follow the plume from a single plant as far as 186 miles from its source in a single day. They found that the sulfur dioxide gas from a power plant was discharged from the stack at a height too high to be removed by structures and vegetation on the ground. Instead, it remained airborne long enough for most of the pollutants to be converted into sulfur trioxide and then quickly to sulfuric acid particles called aerosols. And once the acid aerosols have been formed, the researchers discovered, they can remain aloft for a long time and travel long distances.

The pollution globs, with contributions like these from St. Louis, can stall for days, swirling slowly, collecting more and more sulfur and nitrogen oxides which are converted into still more highly corrosive sulfuric and nitric acids. A third acid—hydrochloric acid—may also be present in significant quantities, especially if metal smelters are adding pollution to the growing miasma. With the formation of these acids, the stage is set for acid rain.

Most of us know the scene well: on a sunny, hot, hazy, humid summer day, thunderheads build up in the afternoon, and the rain falls in the evening. As an acid glob swings through Minnesota on such a summer day, for example, it comes in contact with a cold front moving in from the west. The acid droplets condense into raindrops, and Minnesota is drenched with acid rain.

Summer and early fall, usually from July to September, are the seasons that produce the most acidic precipitation. Scientists believe that there are probably two main reasons for this pattern.

First, the summer is the period in which the generation of electrical power and automobile use—the principal sources of atmospheric acids— are at a peak. Thus thunderstorms occurring during these months of peak pollution tend to bring down the most acid. Secondly, raindrops and, in general, the process by which raindrops are formed, are much more efficient at removing acids from the air than is snow.

Pollution is removed by precipitation in two basic ways, called "rain out" and "wash out." Rain out occurs when a pollutant becomes involved in the precipitation-forming process. Wash out occurs when the pollutant is simply caught by a falling drop. A drop of acid rain, then, can be formed either when a sulfate particle, for example, acts as a nucleus around which a raindrop forms, or when the particle is simply absorbed by a falling drop.

Walt Lyons describes rain out and wash out from a meteorologist's point of view. "One of the strangest things we've seen on these satellite pictures is these large masses of clouds with lots of little holes punched in them," says Lyons. "For a long time we couldn't figure out why there were these holes, but then we realized that what had happened was rain—the smog had been absorbed into the raindrops and fallen out of the blob. We've verified this by comparing precipitation data taken from the ground with the satellite pictures of these holes. What we're seeing is acid rain actually being made."[36]

4. The Scandinavian Connection

Shortly before lunch on a summer day in 1967, the telephone rang in the laboratory of Svante Oden. The young associate professor of soil science at the Agricultural College of Uppsala, Sweden, had recently completed some preliminary research which indicated that the precipitation over parts of southern Scandinavia was becoming increasingly acidic.

On the other end of the line was Ulf Lunden, a fisheries inspector for the city of Uddevalla, about two hundred miles away on the southwest coast of Sweden. Lunden told Oden that he had discovered some lakes in his region in which fish had disappeared and that he had found fish dying off in other lakes. He could find no explanation but had made pH measurements in the affected lakes that seemed far lower than normal. Could there be some connection, Lunden wondered, between Oden's findings of increased acidity in precipitation and the fish kills?

" 'By hell,' I told him, 'It's true,' " Oden recalls.

"A real consequence of the conditions I had found. I was shocked to realize that we were already in a situation where damaging effects could be demonstrated. Lunden made the problem of acid rain a real problem for the first time."[1]

The people closest to a problem—people like Oden and Lunden—are frequently the first to sense that something is wrong, but they have difficulty piecing together the whole picture from their individual vantage points. The flash of recognition between Lunden and Oden followed many years of hunches that something was souring the skies and the waters of Scandinavia.

In the early 1900s, the chief inspector of the fish hatching and catching business in Norway realized that when the baby fish and netted fish were counted in certain parts of the country, the numbers decreased with each passing year. Until 1926, however, he did not know how to explain the decline. "The salmon hatcheries in the southern counties have had great difficulties particularly with mortality among egg and newly hatched fry,"[2] the inspector wrote in his report that year.

"Professor Dahl has suggested that the water acidity could be the reason for the mortality. To test this, I neutralized the water by filtering through lime. The result has been most favourable and at the same time the suspicion has been verified....Mortality increases as the water acidity increases. The only hatchery that has been run with no accident is at Tovdal River. The river here has been found to be naturally neutral. This river has not had the great reduction in salmon fisheries that is found in most rivers in the southern counties."[3]

While the inspector knew the rivers were becoming increasingly acidic, he did not know where the acid was coming from. Finally, in 1959, thirty-three years later, another Norwegian fisheries inspector linked the increasingly sterile rivers with the rain. By that time, however, it was already too late for thousands of lakes and streams in Norway. Even the Tovdal River, in which salmon thrived in 1926, had joined the growing ranks of Norwegian waters devastated by acid rain.

Before long the Swedes discovered that what had happened in Norway was happening in their own country. Scientists who sampled the rain and snow found that all of Scandinavia, especially the southernmost areas, was receiving substantial amounts of acid precipitation.

In 1967, following his fateful telephone call from Ulf Lunden, Svante Oden formulated in detail the theory of how pollution from established centers of industry in Europe can be transformed in the atmosphere into acids that, aided by accidents of weather and topography, fall in Scandinavia where they accumulate in the environment and disrupt aquatic ecosystems.

Five years later, Sweden presented a paper entitled "Air Pollution Across National Boundaries" to the United Nations Conference on the

Environment in Stockholm. This landmark study was the first to focus world attention on the general problem of contaminants drifting across national borders and on acid rain in particular. The growing severity of international pollution was likened to a "form of unpremeditated chemical warfare." Since the 1972 paper, the Scandinavian scientists have continued to uncover most of what the world now knows about the origins and effects of acid rain.

An unfortunate combination of history, geography, and weather patterns coalesced to make Scandinavia the initial focus of the acid rain phenomenon. Heavy industry sprouted and spread and choked the skies of Europe with acid-forming chemicals years before the same thing happened in North America. And the prevailing weather patterns just happened to work in a way that caused much of Europe's pollution to fall on southern Scandinavia.

Scandinavia has been a proving ground for acid rain theories, a training ground for acid rain researchers, and a killing ground for the aquatic flora and fauna that have become acid rain victims. It is, in fact, to Scandinavia that we North Americans should look for a crystal ball view of what we can expect from our own corrosive rains. That view, however, may be grim. Kenneth Hood, who is directing the U.S. government's fledgling effort to assess the potential of acid rain damage in North America, puts it this way:

"We know something about the situation ahead of us. The Scandinavians have been studying acid rain fifteen years longer than we have and their situation has not gotten any better. They've come to the conclusion, for instance, that things are so bad, the damage is so severe, that they may have to lime some of their soils forever if they want to go on using them at all."[4]

While the acid rain crisis in Scandinavia has not given us much hope for the future, at least the Swedish and Norwegian experience has given us important tools to use in making our own decisions on how to counteract the problem. For years Scandinavia was the place that biologists, limnologists, and ecologists from other nations went to find out what acid rain was all about. Gene Likens, George Hendrey, Eville Gorham, Richard Wright, and many others went there to gain firsthand knowledge of acid rain and its damage.

One of the most important lessons these scientists learned from the Scandinavian experience was a surprisingly simple but crucial one: Unless we spend the time and money to study and understand the basic workings of the environment that surrounds us, we will not recognize what is happening when it goes awry. Because the Scandinavians began to take a close look at their environment many years ago, they made the key discovery that acid rain is a pollution problem that transcends local boundaries.

Sweden began studying the chemical contents

of its rain and snow just after World War II by establishing a network of sampling stations throughout the country. The network was extended throughout Scandinavia in the early 1950s and throughout western Europe a few years later. Called the European Atmospheric Chemistry Network, it once included 175 stations, a much more extensive, complete, and organized data gathering system than anything in North America, even today.

The data from this rain collecting network revealed that as early as the 1950s—and probably much earlier—a core of highly acidic precipitation was falling over an area roughly encompassed by Belgium, the Netherlands, and Luxembourg.[5] Svante Oden—sometimes referred to with dubious distinction as the "Father of Acid Rain"—discovered that by 1968 this core of highly acidic rain and snow had become much more widespread and had taken in most of Germany, northern France, the eastern British Isles, and southern Scandinavia. Oden found that the rain in some parts of Scandinavia had become more than two hundred times more acidic during the first two decades of the rain monitoring network. Later, analyses of the network data showed that acid precipitation continued to spread in the 1970s and today encompasses nearly all of northwestern Europe.[6]

Alarmed by the persistent spread of the corrosive precipitation, the Scandinavians wanted to

Changes in the pH of precipitation over northern Europe from 1956 to 1966

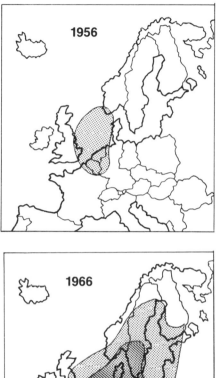

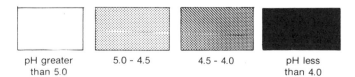

pH greater
than 5.0

5.0 - 4.5

4.5 - 4.0

pH less
than 4.0

know what was causing it. They were sure that human-made pollutants were the source—the grey color from the tars, ash, heavy metals, polychlorinated biphenyls, and other organic pollutants in the snow that fell during acid episodes provided abundant evidence. In addition, they knew that not all the pollution could be coming from their own countries; there were not enough sources of sulfuric and nitric acid in Sweden and Norway to account for the levels of acid in precipitation that were being measured. Oden calculated that at least 70 percent of the sulfur falling on Sweden was coming from outside the country. Again, basic environmental study paid off.[7]

A program of daily airplane flights crisscrossing the skies over northwestern Europe was inaugurated to collect samples of polluted air. When the results were mapped out, the likely sources of acid precipitation became clear. Extremely heavy loads of sulfur and nitrogen pollutants were being boosted into the air over several areas on the European mainland and Great Britain. The principal polluters were in the heavily populated and industrialized centers of London and Glasgow, the Ruhr, Rhine, and Rhone River valleys, Switzerland, the Netherlands, and Czechoslovakia.

Moreover, researchers found that the concentrations of the pollutants released from these areas jumped dramatically between the 1950s and the

1970s—an increase that paralleled the rise in acid rain. In central Europe the concentrations of nitrogen oxides alone increased three to four times from the mid-1950s to the mid-1960s. This rise paralleled the proliferation of industries, oil-fired power plants, and automobiles in Europe after the war.[8] A comparison of the sulfur dioxide emissions from various European countries in the mid-1970s reflects the principal sources of acid-forming pollutants: Norway, 2.5 kilograms of sulfur dioxide per hectare (2.47 acres) per year; Sweden, 6.7; France, 20; Denmark, 30; West Germany, 65; United Kingdom, 131; the Netherlands, 152.[9]

Once the Scandinavians discovered where the airborne pollution was coming from, they wanted to know how sulfur dioxide and nitrogen oxide emissions from a steel plant in West Germany or a power plant in Great Britain could be transformed into acids and fall to the ground in southern Sweden or Norway, hundreds of miles away. The breakthrough came when researchers discovered that sulfur and nitrogen pollutants can remain airborne for as long as four days, more than enough time for the sulfur dioxide and nitrogen oxides to interact chemically with sunlight, moisture, oxidants, and catalysts in a process that produces acidic compounds. Given that these pollutants could stay aloft, all that remained was to test which way the winds blew.

Scandinavian researchers began plotting the

paths of storms that brought rain and snow to their region. Some storms swept in out of the North Atlantic without crossing any other land masses. Others moved first across the British Isles and then into Scandinavia. Still others passed over both Great Britain and western Europe before swinging northward into Sweden and Norway. When the researchers compared the paths of the storms with the amount of acidity in the precipitation that resulted, they found a remarkably simple yet distressing correlation: storm paths that lead through the least contaminated air produce the least acid rain and snow in Scandinavia; paths that lead through the most heavily contaminated air produce the most acidic precipitation.

In demonstrating clearly and for the first time the threat of long-range transport of pollutants, the Scandinavian scientists reshaped the way we think about air pollution. No longer can responsible pollution control officials justify the "out of sight, out of mind" mentality, the false confidence created by the sight of a plume of pollution from a factory or power plant dissipating into apparent nothingness in the sky.

"At a place remote from cities and industries," wrote Svante Oden, "many such tails [plumes] may add up anonymously leading to a change in the chemical climate. This is what has taken place in Europe and a large part of North America."[10]

By the late 1960s, then, Swedish and Norwegian scientists had discovered why their countries were being subjected to rapidly increasing concentrations of acid precipitation. Weather systems passing over northwestern Europe assimilate the pollution "tails" from thousands of smokestacks and tailpipes in Britain, Germany, the Netherlands, and other countries into a thick layer of contaminants. This mass of acid-forming pollutants then passes over Scandinavia where it meets cool air and falls to earth as acid rain and snow—just as pollution from the Ohio River Valley of the United States is scooped up, swept away, and dumped on northern Minnesota and Canada.

The scientists also discovered that the most highly acidic storms occur in Scandinavia only about twenty times a year. Each of these storms, however, can have a serious impact. During one eight-day period, for example, four thousand tons of sulfuric acid were deposited over a section of southern Norway.[11]

The areas subjected to the worst acid drenching are in southwestern Norway, western Sweden, and the interior sections of central and southern Sweden. All these areas have one thing in common; they are made up of mountainous high country which acts like a giant blotter, continuously soaking up the acidic moisture of the storms as they pass over Scandinavia. If these mountains did not exist, the acid-bearing clouds from an air mass that captured the pollution from the power plants of Great Britain might continue on over

Norway without dropping their toxic cargo.[12]

Southern Scandinavia is particularly vulnerable to acid precipitation because of the close relationship between the water in the air and the water on the ground. In Sweden, for example, almost 95 percent of the surface water is a mixture of shallow soil water and groundwater. Thus the acid that is released into the air combines with the transported acid in the rain, and both are added to the lakes, ponds, streams, and rivers. In every period since Scandinavian acid rain research began, a trend of increasing acidity has been recorded in almost all the bodies of water tested, even in otherwise pristine watersheds.

Sadly, the Scandinavians did more than discover the cause and the sources of acid precipitation; they also learned about its devastating effects. In the heavily assaulted southern regions of Norway and Sweden, acid rain damage has been severe. In Sweden alone an estimated twenty thousand lakes have been or are about to be stripped of their fish populations.[13] Thousands more are in the same condition in Norway. In fact, in the lake country of southern Norway, the area where most lakes have lost all their fish is about the size of Switzerland.[14] Swedish researchers predict that if acid rain and snow continue to fall at current rates, 60 percent of all the rivers in the country will reach a lethal level of acidity within twenty-five years, and 90 percent will attain that lethal threshold by the year 2050.[15]

Confronted by this serious threat, the Scandinavians have led the way in documenting the extent of the damage that has already been done and in anticipating the damage that may occur in the future. For example, Scandinavian scientists were the first to determine that when the pH of the rain falling into a vulnerable lake reaches a threshold of 4.6, fish usually begin to disappear. American scientists have since observed a similar threshold in acid-damaged lakes in such areas of the United States as the Adirondacks and the Boundary Waters region of northern Minnesota.[16]

Swedish scientists were the first to note a bizarre effect of acid rain in which sphagnum moss, a plant that normally grows on land, invades the bottom of acidified lakes and chokes out other plant growth. The same phenomenon has been observed in lakes in the Adirondack Mountains. In both places the takeover by the sphagnum moss hastens the disappearance of fish and, ultimately, the demise of the lake.

Swedish researchers were also the first to document the steps through which a lake passes as it becomes acidic. They demonstrated that a lake can appear nearly normal as long as its neutralizing substances are capable of providing a buffer against excess acid. Then, when the neutralizers are exhausted, the lake suddenly becomes an inhospitable mix of dilute sulfuric and nitric acid.

Working together, Finnish and Swedish scientists were able to further refine their knowledge of this acidification process. Comparing data from a systematic study of water quality in lakes and rivers in both countries and data from studies of acid precipitation in those areas, they were able to determine how long it took for those waters to become acidic in the presence of a given amount of acid. Such pioneering research will play an important role in the United States when and if government officials consider setting new limits on the amounts of acid-forming pollutants that can be released into the air of each region.

The impact of acid precipitation in Scandinavia may be most serious in the acidification of soils. In southern Scandinavia, where farming and forestry are both economically important activities, researchers are now concentrating their efforts on learning how acid rain affects soil fertility. Because the soils in the regions that receive the most acid precipitation are thin and already slightly acidic, their capacity to absorb much more acid is limited. The most severe soil acidification has occurred in a continuous band in southwestern Scandinavia that is also the area closest to the source of acid-forming emissions in Great Britain and Europe.

In order to establish that acidified soil conditions in southern Scandinavia are the result of human activities, Swedish scientists conducted an ingenious study that analyzed the soil in the area surrounding a copper smelter near Falun that has been operating for about five hundred years. In those five centuries, the smelter discharged an estimated 5 million tons of sulfur dioxide. The study showed that the soil close to the smelter is more acidic and contains fewer nutrients than the soil farther away from the smelter.[17]

Though Scandinavian scientists know that their soils are becoming increasingly acidic and that acidification has resulted in a loss of soil nutrients, they still do not know how acid precipitation will affect plant growth. Laboratory experiments have shown reduced growth and a reduced success rate for the seedlings of some plants and trees, but acid stunting has never been conclusively demonstrated in the field. The impact of acid on plant growth is one of the most important areas left to be resolved in both Scandinavian and North American acid rain research.

While the effects of acid on plant growth are uncertain, the devastation caused by acid rain in the aquatic ecosystems of Scandinavia has been dramatically demonstrated in southern Norway. In the spring of 1975, thousands of fish suddenly died in the Tovdal River. The incident was intensively investigated by a special team of scientists that had been assembled by the Norwegian government and scientific community in the early 1970s to determine the impact of acid rain on forests and fish. The scientific team not only linked the sudden death of the fish to acidic snowmelt, but it also compiled evidence of a pattern of destruction in the watershed's ecosystem that has taken place over many decades—a pattern that has been found repeatedly throughout Scandinavia.

The Norwegian fisheries inspector who gave the Tovdal River a clean bill of health in the 1920s spoke too soon, for even then the river was being degraded by the acid accumulating in its watershed. The Tovdal is a major river in southern Norway that drains a 680-square-mile area. Most of this area is covered by thin soils underlain by granite which is highly resistant to chemical weathering—terrain similar to the acid-sensitive areas in North America. Coniferous forests, peatlands, and rock barrens cover much of the land, and not many people live there because there are few industries and little farmland. Like much of southern Scandinavia, however, it projects a wilderness aura of rugged beauty.

The Tovdal was once one of the major salmon-producing rivers of Scandinavia. When the fish traveled up the river to spawn, they were netted by commercial fishermen. Records of catches from the Tovdal have been kept since 1880, and at the turn of the century seven tons of salmon were hauled from the river each year. From that time on the salmon catch dropped steadily, and since 1970 it has been nil.[18] In fact, salmon industry records show that the catch from all seven major rivers in southern Norway had dropped to zero by the 1970s. The salmon catch in other Norwegian rivers, however, has shown no such precipitous drop.[19]

The Tovdal and the other rivers ravaged by acid

in Norway are in the area that is most heavily
saturated with acid precipitation. Scientists estimate that the basin receives from five to eleven
tons of excess sulfuric acid per square mile each
year. The lower end of the river near the coast
receives from two to three times as much acid as
the headwaters.[20]

Although scattered observations were made of
the acidity and chemical content of the Tovdal
River water starting in the 1920s, regular sampling
did not begin until 1968. In the 1920s inspector
Sunde characterized the Tovdal as a "neutral"
river; today the situation has changed dramatically. The pH plunges regularly below the critical
threshold, fluctuating between its most acidic
level during spring snowmelt and autumn rains
and its least acidic level during the summer. Such
conditions caused thousands of fish to die in the
Tovdal in the spring of 1975.[21]

In the Tovdal watershed, trout and perch used
to be the most common fish species. Trout were
once found in almost every lake and stream, and
sport trout fishing was a socially and economically
important activity in the region. Now the trout
have disappeared from most of the waters in the
Tovdal basin in a decline that has been most rapid
during the last twenty years. The fish remains only
in the main river and in a few large lowland lakes
that have not yet become intolerably acidic.

Of the 266 lakes in the basin surveyed by
Norwegian scientists, 171 are without fish, and

100 have lost all their trout in just the last fifteen years. Of the 95 lakes that still contain fish, more than 50 have sparse and declining populations.[22]

Despite meticulous research elucidating vivid cases of acid damage to aquatic ecosystems, Scandinavians still debate the causes and effects of acid precipitation. Svante Oden, the pioneer acid rain researcher, described the debate in these terms. "Attitudes are highly polarized from almost a total denial of the problem to my [position on] the impact of acid rain. This arises from the fact that short-term economic gains achieved by pollution of air or water stands against the cost of long-term deterioration of the environment."[23]

In 1974 a group of scientists met in Winnipeg, Manitoba, to discuss acid precipitation and other problems affecting freshwater ecosystems. Oden was one of those who adopted the following resolution:

"Whereas the increased introduction of man-made pollutants to the atmosphere is seriously contaminating the earth's airsheds, often remote from local sources, and

"Whereas the fallout of these materials is contributing to acidification and other pollution of lakes, rivers, and ground waters of large geographic regions, and,

"Whereas the recently observed and projected changes in acidity of waters represent a serious stress for natural aquatic ecosystems,

"Therefore this congress deplores such degra-

As consumption of fossil fuels has increased in Europe . . .

Coal and oil consumption, millions of tons

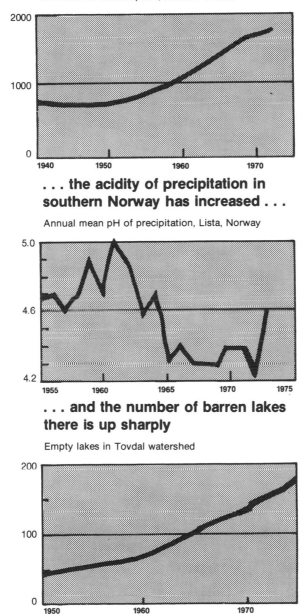

. . . the acidity of precipitation in southern Norway has increased . . .

Annual mean pH of precipitation, Lista, Norway

. . . and the number of barren lakes there is up sharply

Empty lakes in Tovdal watershed

Source. "Impact of Acid Precipitation on Forest and Freshwater Ecosystems in Norway," SNSF Project, Research Report FR 6/76

dation of aquatic ecosystems and urges governments, scientists, engineers, and laymen everywhere to investigate thoroughly the ecological
magnitude of these changes and to undertake
prompt and ecologically sound remedial action."[24]

Svante Oden believes, however, that such a
remedy may be a long time coming. "There is no
agreement as regards either the amounts or the
effects of this international transport of substances," says Oden. "As long as this state of
matter persists, reductions of [acid] emissions are
not likely to be made."[25]

5. *The Acid Rain Pioneers*

This is the man who started it all, the man who first told America that the rain had turned to acid. Here in his drab college office in Ithaca, New York, Gene Likens of Cornell University sits among books and journals and charts and files and tables and endless columns of numbers—saying he's sorry he doesn't have more numbers.

"Scientifically, acid rain is not a very satisfying field of study," says Likens. "I like to do data intensive research. I like to compile massive amounts of data. In acid rain, the data is very difficult to come by.

"I like hard science, the science you can test. You ask a question, test the question, and form a conclusion. I like that to be my guiding light, but emotional issues like acid rain do pop up."

Twenty years ago Likens, a scientist's scientist, was contentedly and quietly hard at work studying the life cycle of a small New Hampshire stream. Then one day he happened to take a closer look at the rain. Ten years later, he would bring the

American people some very bad news and, in doing so, he would leave behind permanently the comfortable obscurity of the laboratory.

Gene Likens is one of a handful of scientists in the United States and other nations who have found themselves in the difficult position of being the first to decipher some aspect of the acid rain puzzle. Among them are Eville Gorham, a Canadian who made some of the earliest studies of acid rain; William Lewis, a University of Colorado scientist who found acid rain in a part of North America where it had not been expected; Svante Oden, a Swedish researcher who made the connection between acid in the sky and death in the lakes and launched an international crusade for corrective action.

While these pioneering scientists have experienced the joy of breakthrough, they also have had to endure the natural skepticism and contrived criticism that attends most discoveries. They have seen their work attacked publicly in journals; they have seen other scientists hired by affected interests in an attempt to discredit their work; they have seen sources of money to continue their work become scarce; and they have had to cope with the jealous scorn of colleagues. Still, they persevere in their research and remain convinced that the danger from acid rain is real.

In the early 1960s, Gene Likens and F. Herbert Bormann of Yale University were studying the biology, chemistry, and precipitation of an area in

New Hampshire called Hubbard Brook. They were trying to measure everything that entered that ecosystem, everything that came out, and to describe what happened in between. Likens and Bormann produced volumes of numbers, and among them were measurements of the acidity of the rain falling on Hubbard Brook. "It was obvious that the rain was acidic but we just assumed it was a local effect," Likens remembers.

Likens knew that Scandinavian scientists were probing a then scarcely understood phenomenon called acid rain in Sweden and Norway. But until he went to Sweden on a fellowship and talked to Svante Oden and other researchers there in the late 1960s, he did not realize that the acid rain problems of New Hampshire and Sweden were the same. As a result of this trip, says Likens, "I began to wonder if acid rain might be a regional thing. So in 1969 when I moved from Dartmouth to Cornell, I started making measurements in central New York state. I was surprised to find the same kind of acid rain data that I'd been getting in New Hampshire."

In 1972 Likens—with Bormann of Yale and Noye M. Johnson of Dartmouth—published a landmark paper that warned for the first time of the acid rain danger over the entire northeastern United States. "I had thought this earlier but I wasn't ready to put it on paper until later," says Likens.

Likens's reluctance to go public with the acid

rain research is understandable for two reasons. First, he is a reticent, deliberately understated man who must be dragged into speculating on the implications of his work. He composes each word, sentence, and thought to say grudgingly exactly what he means and no more. For instance, "If you pour acid on living things, they will be stressed;" not "killed," "annihilated," or "destroyed," but "stressed."

Likens also understood the implications of research that placed blame on such a powerful sector of the economy as the electric power industry. The 1972 article, and others that followed, made him—as well as acid rain—a major controversy. His research was no longer just a matter of scientific curiosity. Likens himself became a target of professional jealousy and the focus of opposition by the power industry. One prominent utility research group even "hired two Ph.D.s to discredit what I've written," Likens says.

In the mid-1970s, the assault on Likens's work was carried out in the pages of the widely read journal *Science.* Two scientists wrote a paper contending that sulfuric and nitric acid were not the main acids present in acid rain and, consequently, power plants and other large pollution sources could not be responsible. One of the scientists worked for the Federal Power Commission, a now defunct federal government agency that dealt with electric utilities.

The critics were rebuffed by a group of scien tists who supported Likens's claim that sulfuric and nitric acids are the main acids in acid precipi- tation and that they do come from power plants and any other place where fossil fuels are burned. In 1976 Likens and his colleagues wrote another paper buttressing their earlier work and dismissing their critics. Today he is certain that the electric power industry is the prime source of acid rain.

No study has yet established that a molecule of sulfur dioxide from a power plant in Ohio or Illinois, for instance, ends up as a droplet of sulfuric acid on a soybean leaf in Minnesota. Likens, however, believes that the weight of circumstantial evidence implicating power plants is overwhelming:

● Power plants discharge the largest share of the sulfur dioxide and nitrogen oxide pollution in the United States.

● The increase in the acidity of rain and snow in the United States during the last thirty years paralleled the rise in the discharge of sulfur dioxide and nitrogen oxides by power plants.

● Power plant and industrial smokestacks have been built much taller since 1950, enabling the gaseous precursors of acid precipitation to travel much farther.

● At the same time that smokestacks were getting taller, pollution controls were being installed that removed large soot particles from smoke. The soot particles, however, neutralized acids formed in the air and caused them to fall to the ground near the pollution source. Thus the pollution controls freed the acids to drift hundreds of miles from their source before they fell to do damage.

● Scientists have used aircraft-borne instruments to track a cloud of sulfur dioxide and nitrogen oxide pollutants from a single power plant over long distances.

● The huge masses of sulfur and nitrogen oxides that form in the atmosphere before acid storms have been shown to develop first over the areas where there is a large concentration of power plants and coal-burning industries, as in the Midwest, or a large concentration of cars, as in the Los Angeles basin.

● The acidity of rain and snow is highest beneath the paths followed by the masses of pollution that originated in the areas where power plants are concentrated.

Likens's conclusions about the origins of acid rain are now almost universally accepted in the scientific community, and he is regarded as one of the principal figures in acid rain research. While he believes that publishing articles about acid rain is necessary to create an awareness of the problem and to prompt others to investigate it, he is clearly uncomfortable with the role of public scientist.

"I find notoriety disruptive and disturbing. I don't like it. I don't like it at all," says Likens. "Every time an article appears, I get calls from all over the world. Crank calls blaming me for much of the world's trouble. After a 1974 article in the *New York Times,* I had to go and hide. For a conservative scientist it was a shock."

Squirm as he might about being caught in the crossfire between environmentalists and polluters ("I'm an aquatic ecologist not an environmentalist"), Likens considers the public exposure and controversy part of the job of a responsible scientist. "I don't regret this phase. It's an emotional one. It's emotional because, from the very start, you are talking about the rain. People can't imagine there is anything wrong with the rain; it is something everybody experiences."

It is also an emotional issue, Likens says, because "it involves big money. Industry is trying to protect its interests. I don't like that part. I don't think the resolution of all problems should depend on economics. But unfortunately the way we operate in this country is that nobody spends much to clean up air pollution unless there is an economic reason."

Ironically, one of the major results of the process that Likens set in motion in 1972 is that the government now sees acid rain as an economic problem and has begun spending research money to fill in some of the gaps in knowledge about it. Late in the summer of 1979, President Carter called acid rain one of the top environmental concerns of the United States and ordered $10 million spent in the first year of a ten-year research program on its causes, effects, and possible solutions.

Likens says the acid rain situation will continue to cause controversy as many more scientists begin offering their views, as more reports on the problem appear in print and on radio and television, and as industries try to avoid being stuck with the blame. Though Likens would have preferred to remove himself from the fray, he served on a committee that decided how the acid rain research ordered by Carter would be carried out. He saw it as his chance to make sure that the money is spent on "hard science."

"We've described the problem," says Likens. "Now it's time for more study of the causes and effects. I have to be very guarded in recommending specific solutions to problems. I can tell you that the more sulfur dioxide and the more nitrogen oxides you put in the air, the more acid rain you'll have. How much more I can't tell you yet. If I'm wrong, I'll be delighted. If I'm right, the problem is very serious and will have to be dealt with."

"I'm getting tired of acid rain. If people twist my arm, then I'll go talk about it. But I've been talking about it for a long time and now acid rain is a mainstream issue. I don't like being in the mainstream of ecological research. I like to be on the fringes."

Eville Gorham was indeed on the fringes of things twenty-five years ago when he was trekking through the peat bogs of northern England collecting raindrops. But what he found in those drops would, over the next two decades, make the work of the young Canadian the foundation of one of the most important environmental issues in the Northern Hemisphere and make himself a sought-after expert on acid rain. In recent years, it has been difficult to attend a meeting, symposium, or conference on. acid rain where the ubiquitous figure with the black-framed glasses, combed-back hair, serious visage, and blended British/ Canadian accent couldn't be found behind a podium offering his measured, scholarly recitation of the facts.

In the early 1950s, though, Eville Gorham was a postdoctorate student in Sweden studying some routine work by Swedish scientists on how the rains affect the life of ecosystems. Afterward he went back to England, where he had received his Ph.D. in botany, to follow up on what he had learned in Sweden. He set up his research project in the lake district of northwestern England, a remotely undeveloped area of mountain lakes and bogs. "The first thing I found," says Gorham, "is that we were alternately drenched by rains dominated by sea salt when winds blew from the Irish Sea to the west; and if the wind was blowing from the south, from industrial Lancastershire, or from the east, from North Umberland, we were getting drenched with sulfuric acid.

"I tried to work out the effect of daily weather conditions—how much rain fell and which direction the wind was coming from and so on. And of course it was fairly easy making the connection between acid in the rain and industry because when I got sulfuric acid, the filters through which I was collecting my rain samples were sooty as all get out."

Although Gorham didn't know it at the time, other researchers in America and Sweden were stumbling upon acid rain in rural areas far from factories and cities. In fact, researchers had been recording pH measurements of rain since about 1939. And yet the realization that acid could escape from industrial areas into rural areas did not occur until 1955 when three independent studies were published, one by Gorham and two by other researchers.

"Nobody had the idea that they were studying air pollution," Gorham remembers. "It was all just serendipitous; I was interested in the nutrition of oligotrophic ecosystems. Barret and Bodin in Sweden were working meteorologists tracing air masses. Haughton in America was just a general meteorologist; he wasn't particularly interested in acid rain. He was studying clouds, and it just turned out that he found it.

"It was all accidental as so much research is. And of these groups I was the only one who really carried on with it. I recognized it and really pursued it. I started looking at it in more detail. I really wanted to get at the ecological effects, but I didn't know how. We didn't have any good data, and there were no organisms where we knew much about pH tolerance. So I thought the most interesting thing I could do to show an effect on biota would be to look at effects of sulfuric acid on mortality of respiratory diseases in British cities."

In the late 1950s, Gorham compared the incidence of bronchitis, pneumonia, and lung cancer in an English locale with data gathered by another researcher on the amount of various air pollutants that fell there. The results were disturbing. Gorham discovered that increasing amounts of acidic pollutants produced a greater incidence of bronchitis. The more sulfate particles in the air, Gorham found, the more pneumonia, and the more tar, the more lung cancer.

Intrigued by the effect of acid on living things, Gorham shifted his research focus into the distant past. He decided to take drill-core samples of sediment layers in the earth and locate fossilized remains of microscopic life forms in order to learn how the array of life in the area had changed with the advent of acid rain.

When moving the bulky drilling equipment into the English lake country proved to be impracticable, Gorham took his research back to Ontario in his native Canada. There, not far from the notoriously polluting copper and nickel smelter at Sudbury, he expanded his acid rain work to include a study of the effect of sulfur pollution on plant communities and aquatic life.

Despite his pioneering studies, Gorham has never thought of himself as an "acid rain" scientist, and his work has led him into other areas. "I've spent my life doing pollution ecology, but I've always thought of myself as a fundamental ecologist, not a pollution specialist. It just happens that I've been interested in acid rain. I knew damn well [by the early 1960s] that it had an effect on the biota, but after a while I didn't pursue it."

Gorham moved to the University of Minnesota in 1962 and involved himself in other ecological studies until he became aware of the work of those scientists who were studying acid rain elsewhere— work that shed new light on the immensity of the problem. "I hadn't thought of acid rain as traveling a thousand kilometers," says Gorham. "I had seen in England and Ontario that some acid could go one or two hundred miles, but I didn't have any idea of the thousand mile distances Svante Oden found in Sweden. Why did it take so long for us to realize how widespread the problem was? I guess it takes a certain amount of time for effects to become so overwhelming that people can't escape them."

Gorham went back into direct acid rain research when he realized that Minnesota offered a rare opportunity. "After reading the articles that Gene Likens of Cornell and his friends had been producing, it was clear that the problem was spreading. And it appeared that Minnesota was at an intersection, so to speak.

"It is exposed to acid rain from the east and, on the other hand, it's next door to the cultivated prairies with a lot of dust blowing in, alkaline dust that can neutralize acid. Minnesota seemed to be a critical area for the problem, a tension zone between the acid from the east and the alkali from the west.

Unfortunately, Gorham's idea fell on deaf ears—a frustrating experience that has plagued other would-be acid rain researchers. "I tried to get some funding from the National Science Foundation twice in a row without success," he remembers. "I put in what I thought was an ecological proposal to trace the sources of the materials entering ecosystems. I wasn't interested in atmospheric chemistry. I was interested in ecosystems, but I couldn't get the ecologists to look at it. The atmospheric people rejected it because they said the field had passed me by." Even a recommendation from Likens couldn't break the bureaucratic barrier, and Gorham's proposal stayed on the shelf for months until the Energy Research and Development Administration decided to approve funding for it.

To begin the project, Gorham chose three sites at which to sample rain and snow. The sites formed a line that ran from the forested Boundary Waters region of northern Minnesota, through the transition zone between forest and prairie in central

Minnesota, and into the open prairie and agricultural area of eastern South Dakota.

The samples revealed that the acidity of the rain and snow increased the farther away from the plains—and the alkaline dust—one went. In the Boundary Waters region, a wilderness country of pine forests, granite outcroppings, and fish-filled lakes, he found that the average pH of the precipitation was an ominous 4.66. "This is just in the intermediate zone between pH 4.8 where the Swedes say they don't see major ecological effects and 4.5 where they see striking effects," says Gorham. "The amount of sulfate falling on this region is close to 15 kilograms per hectare per year which is where the Swedes have seen a rapid increase in the acidity of sensitive lakes. So it seems reasonable to me [to conclude] that in the Boundary Waters we're just on the edge of trouble. If Gene Likens is right and the problem is expanding, worsening—and I think he is right— then it's going to get worse in Minnesota.

"We already are seeing damage in the aquatic ecosystem—the deterioration of lakes in the Adirondacks, the fact that lakes are being acidified in Maine and Nova Scotia. If we wait to demonstrate beyond the shadow of a doubt the scientific questions, the damage will have been done. The longer we wait, the more difficult it will be to get moving."

While Gorham was using his research to expand scientific awareness of acid rain, he was using his sonorous speaking voice and skill with groups to share his warning of impending destruction, often to the distress of more reserved colleagues. "How many times have I spoken? So many, it's hard to know," he says. "I've given hundreds of talks in the last fifteen years to any group you care to name: church groups, school groups, the national guard, legion posts, 3M. It takes a lot of time, but I feed at the public trough and I reckon it's my duty as a member of this institution [University of Minnesota] to go out and do my bit to educate the public. I suspect that's more important than the classroom teaching I do, quite frankly."

Gorham is surprised by the numbers of people who gather to hear his grim words—close to one thousand in Toronto in the fall of 1979 and more than five hundred on a winter Saturday in Minneapolis. He welcomes the interest because he believes that "public attention on the issue can do nothing but good."

While the alarm bell tone of many of these public meetings, talks, and articles about acid rain often disturb an eminent researcher like Gene Likens of Cornell, Gorham has a more pragmatic disposition: "I have a point of view on this issue that is perhaps a little unusual among academics. My view is that politicians don't get elected by cool, calm, rational argument. Advertisers don't sell products by cool, calm, rational argument. And you don't get an environmental issue before the public with cool, calm, rational argument.

"There's got to be some hyberbole, some painting of black and white," argues Gorham. "I reckon the other side is going to paint the whole issue white, so I paint the black side and I expect this may get the issue out before the public where sooner or later there'll be hearings of fact before an impartial arbiter where cool, calm, rational arguments can take place.

"I prefer not to paint things like that myself. I try to avoid hyberbole when I can but I do admit to a definite bias. I think it's a hard line to walk: You have to live with yourself, but on the other hand you have to work hard to get the issues out where there will be some discussion," Gorham emphasizes.

"I've been criticized by some colleagues for this. But I think universities are self-selecting for people who want to opt out of the real world. It's a nice quiet haven, and those of us who do raise issues in the real world are a threat to those people who want to stay safe. If we're doing it and we're right, then they're wrong and they ought to be out there."

Once every week for about two years, someone from William Lewis's laboratory in Boulder, Colorado, had made the forty-five-minute drive up tortuous canyon roads to the small mountain clearing along Como Creek, just a few miles east of the

the Continental Divide. Atop a ten-foot wooden tower in the clearing sat a sophisticated bucket which was designed to catch the rain and snow that fell each week. Usually technician Ronna Edgett would collect the week's worth of water, drive back to Boulder, perform a routine set of tests to see what the precipitation was made of, log the results in a computer, and wait a week before repeating the same process.

Lewis, an aquatic biologist at the University of Colorado, did not look closely at the computerized data during those first months; he considered the data to be merely the groundwork for a larger study he and colleague Michael Grant were preparing of the life cycle of a small watershed in an undisturbed wilderness. Imagine his surprise, then, on the day in 1976 when he happened to glance at the numbers.

Far from reflecting an unspoiled wilderness, the measurements of acidity on the computer printout suggested a polluted city. Almost 90 percent of the readings were more acidic than natural rain and snow, and many were as acidic as readings found commonly in heavily polluted regions. When Lewis compared the readings from year to year, he discovered that there was an unmistakable trend to increasingly acidic precipitation.

Imbued with new importance, the experiment was carried on and has continued to reveal the presence of more and more acid falling on waters in the high reaches of the nation's treasure, the Rocky Mountains. Acid rain and snow were no longer the problem of the industrialized East; they had invaded a pristine high mountain wilderness where soils are thin, the weather extreme, and nature cannot tolerate much disruption.

In the shaded backyard of his small, one-story home in the outskirts of Boulder, William Lewis tells a story of his frustrated quest to pinpoint the source of the acid falling in the Rocky Mountains. It is a story told by a man whose brand of science is all business. Lewis is the kind of scientist who puts up a memo on the laboratory bulletin board informing his students, assistants, and technicians that the only conversation allowed in the lab is that strictly related to laboratory work. He works slowly, deliberately, meticulously. Those who work for him view him with respect and admiration.

Much of Lewis's research training has involved devising ways to study phenomena for which data gathering was a challenge. At the University of Georgia, for instance, he wanted to study the ecological impact of the common practice in the South of large-scale, deliberate burning of forests and brush. To carry out the research, he built a chamber in which he recreated miniature forest fires, trapped the gases and particles released, and analyzed the contents.

When Lewis came to Boulder in 1974 and decided to study precipitation at elevations above

9,000 feet, he faced a similarly difficult proposition. The heavy snows in the Como Creek watershed are sometimes preceded by hail or sleet or both, mixed with rain, or followed by brilliant sunlight that can sublimate moisture rapidly. Yet Lewis and Grant came up with a precipitation-collecting apparatus that can collect and melt snow into easily preserved water as fast as it falls.

While for Lewis overcoming the technical obstacles to research projects has been a difficult task, dealing with the reaction to his findings has been even harder. He acknowledges the controversy that is created by research directly related to the production of energy in an energy-short nation. And even though he was inspired in his acid rain study by the early research papers of Eville Gorham, he does not share Gorham's willingness to align himself with a cause. Like Gene Likens, Lewis would rather just carry out his research and make the results known; he has little tolerance for headlines or politics.

"Our findings were very controversial," Lewis remembers. "We had calls from representatives of Denver saying this could be damaging to the image of the area. We had calls from housewives alarmed and wondering what to do to avoid the rain. But what do we do? Sit on the information?"

In contrast to the strong public reaction to Lewis's research findings, the scientific community showed a distinct lack of interest. Ironically, Lewis's lab in Boulder is just minutes away from the National Center for Atmospheric Research (NCAR) where a sophisticated group of weather researchers fly airplanes high into the atmosphere to carry out experiments. The NCAR scientists did not appear to be pleased with the revelations from Lewis and Grant.

"They tended to dismiss the findings as just a result of pollution from Denver, but I think they were a little disturbed that they hadn't known about it," Lewis says.

NCAR reluctantly agreed to compare its weather records with the Lewis and Grant acid data. When the comparison failed to pinpoint Denver as the prime source of the acid, however, NCAR wasn't interested in pursuing the matter, Lewis says. "This is a problem that runs through the whole acid rain controversy. NCAR and its atmospheric scientists just are not interested in what hits the ground," Lewis claims. "Their science doesn't deal with it. They study the air, the particles, the light and don't make the connection with what comes down in the end. That's why all of the early work on acid rain has been done by aquatic biologists. At some point there has to be a marriage between the two disciplines but it hasn't happened here."

Lewis and Grant have been continually frustrated in their attempts to obtain funding for additional research to find out how widespread acid rain is in the Rocky Mountains and where it comes from. "We felt we should answer the questions that people, especially people in Colorado, want answered," says Lewis. "There should be great interest because the mountains are a high-risk area as far as acid rain goes and because of the degree to which we depend on them economically and aesthetically."

Lewis wanted to establish a network of rain and snow sampling stations like the one at Como Creek in a larger area of the Rockies. His relatively modest proposal envisioned setting up about twenty stations and using forest service rangers to collect the samples and the laboratory in Boulder to analyze them. This two-year project, which would require at most $150,000 a year, was turned down by those in charge of allocating the acid rain research money in Washington.

Lewis and Grant even testified before Congress, enlisted the aid of Colorado senator Gary Hart, and made personal appeals to the Environmental Protection Agency for money. The only funding they could obtain was about $25,000 to develop a method of sampling and analyzing snow; no money was available to determine how widespread acid precipitation had become in the Rockies.

"Mike and I have essentially given up," Lewis says sadly. "We just can't break this barrier. The University of Colorado has sustained our work, but without other money we can't afford to run the lab. We're just going to stop trying."

Svante Oden says it's time for a second look at acid rain. A decade ago he told the world in a report to the United Nations that the continuing acidification of the global environment by industrialized nations was tantamount to chemical warfare. He was the first to see clearly the link between the stricken, fishless lakes and rivers of his native Sweden and the growing acidity of the rain and snow. Oden was also the first to call for immediate international action to deal with the newly recognized problem.

Today, at age 47, Oden is still at the forefront of his field and still saying that something must be done. A man of quiet, understated humor, he pursues his work on the effects of acid precipitation on soils as a full professor at the Swedish Agricultural University in Uppsala, northwest of Stockholm. He is a consultant to the Swedish government and recently completed a report bringing the scientific community up to date on the seriousness of the acid rain threat, a report Oden calls a second look.

"Some things were right, and some were wrong ten years ago when we started," he says. "We have to change our view somewhat. Acid rain is not a scientific problem for stubborn people."

Oden says that in the United Nations report he was farthest off the mark in predicting that acid rain posed an immediate and serious threat to forest growth. "It was my fault, my computation. I said that acid rain would retard forest growth on the order of one-half to one percent a year. That turns out not to be true." Oden had thought, and many scientists had agreed, that the ability of acid to make such nutrients as calcium less available to tree roots would slow growth.

"What we have found since," says Oden, "is that growth is much more closely related to the amount of nitrogen and phosphorus present, and there has been a doubling of these elements [in Sweden] in the last ten years due to increased energy production in the North Sea area [resulting in increased nitrogen oxide deposition] and increased use of fertilizers." While his original theory about how acid might harm forests has not been confirmed, other work now underway in Germany indicates that the salubrious effect of nitrogen and phosphorus might be offset by the ability of acid to slowly but surely kill the root systems of trees. According to Oden, the resolution of this issue is yet to come.

Oden now admits that he underestimated two critically important parts of the acid rain equation: first, the speed at which the environment is being acidified; second, the lethal role of toxic metals. A decade ago, Oden says, "I made the estimate that with current loadings of acid in precipitation, it would take 50 to 150 years for the pH of water systems to decrease to a destructive level with respect to fish populations. We now have seen that the speed of acidification is much faster, 10 to 20 years and sometimes faster."

In addition to finding that lakes and streams are becoming acidified more quickly than he had previously estimated, Oden now realizes that such metals as aluminum and mercury are primarily responsible for the death of fish and other aquatic creatures in acidified regions. And it is this realization of the central, deadly role of metals that prompts Oden to say that the threat of acid rain is now more serious than when he began studying it. "When I started research, metals were only presumed to be a threat but there was no real study going on. In the 1970s, though, it became clear that the fish kill is a synergistic combination of acid and metals. These are the keys and there are no new elements that will come into the picture."

Svante Oden has experienced some of the same frustrations that other acid rain researchers have confronted. Soon after Oden had made the connection between data he had compiled on the acidity of precipitation and fish kills reported by a fisheries inspector on Sweden's west coast, he found that few scientists were willing to listen to him, let alone encourage him to continue his research. "When I recognized the consequences of acidification," he remembers, "I was very interested in pursuing the research further. But others argued against my doing so." One of these scientists was a close colleague of Oden, the head of the soil science department at the agricultural college, and the man for whom Oden worked. "He told me that acid rain research was rubbish and nonsense. It is not possible, he told me, to measure pH in precipitation. Almost all the results are fake, he said. He went for fifteen years without any comment to me on the subject of my work."

During those years, Oden not only pursued his research, but also achieved acclaim as one of the preeminent acid rain scientists. In the early 1960s he was instrumental in establishing a network of observation stations in Europe to measure the acidity of lakes, rivers, and streams. This network complemented an existing network of precipitation monitoring stations and has provided the kind of data most needed in acid rain research—data relating the amount and rate of acidification of lakes, rivers, and streams to the acid strength in precipitation. "We can see what goes in and what comes out and how ecosystems react," Oden says. That kind of information eventually will allow scientists to construct a model predicting what kind of effects a certain amount of acid-forming pollution will have in a certain region.

Until that kind of model can be constructed, Oden says, nations polluting the rain must begin taking steps to control acid-forming emissions. During the ten years since the United Nations report, however, he has seen little evidence that his warnings have been heeded. "When I came up with these ideas, I was convinced that the impact of acidification would make nations act firmly and quickly," says Oden. "I said then that there was a chemical war going on in Europe. We had at that time enough information to start doing something to stop it, but nothing happened. The degree of frustration I have felt in my career relates mostly to all those people who are arguing all the time against taking any action at all.

"The problem of acid rain is economically enormous. It involves the whole structure of society. We have been living with a philosophy of emitting all the time and now see dimensions of the damage. It is time for a whole new way of thinking about energy production. So I can understand why politicians hesitate to start, but it is definitely not too late to do so."

Oden says that those in the United States and Europe who argue that we can afford loss of some lakes and animal species to acid rain are wrong. "It is impossible to put a dollar value on the damage to fish and water life and compare that to the cost of taking out the sulfur. People shouldn't live only for the economy. They should also live for the fun of living; otherwise we should all just quit. What is the fun of living if we are just machines in some economic system?"

Even though there is more discussion today within and among nations about what to do about acid rain, Oden isn't optimistic about the results of such talks. "There was a meeting in Geneva last autumn [1981], a meeting of thirty nations. Only two of them—Sweden and Norway— presented any clear plan for sulfur emission reductions. Most had no plans for any reduction and some said

they would increase emissions. So it doesn't look
good. I'm more optimistic about the United States
and North America. There are only two countries
involved, you and Canada. Within Scandinavia,
we have made progress with respect to emissions
by solving it on a three-nation basis. It's much
easier to reduce trans-border effects when there
are only two or three of you doing the talking."

6. A Test of American Will

Room 2123, Rayburn House Office Building, Washington, D.C., 1:20 in the afternoon of February 27, 1980. The august hearing room with its vaulted ceilings, green velvet curtains, and walnut-paneled walls was packed for the second day of conflicting testimony. Television lights and cameras fixed on Bob Eckhardt, the representative from Texas who was presiding over this first crucial public fight about acid rain.

In one corner was Douglas Costle, head of the U.S. Environmental Protection Agency, flanked by a witness list of those who were sounding the acid rain alarm. In the other was Lynn Coleman, top lawyer for the Department of Energy, flanked by those who said the alarm should be stilled, at least for the time being, in the national interest. And with Eckhardt in the judges box were members of Congress who would decide how to lead the country in this first test of resolve on a most difficult issue.

The discussions, debates, and deliberations

that would take place in this room and in other
similar forums on Capitol Hill during the spring
and summer of 1980 would recreate in microcosm
the controversy over the existence and implications of acid rain that had been building for a
decade. They would also provide a vivid illus-
tration of the enormous political and economic
obstacles that have to be overcome before the
United States adopts a policy of controlling the
pollutants that cause acid rain.

The year 1979 had been one of awakening to
the meaning of the words, acid rain. By year's end,
however, the issue was still mainly that—a matter
of words. A commission had been formed here, a
study group formed there, and monitoring plans
proposed. Still, as Robert Rauch of the Environmental Defense Fund said in the spring of 1980,
"The amount of action on acid rain is inversely
proportionate to the amount of rhetoric."[1]

In 1980 the acid rain controversy began in
earnest. Warning calls for a national policy to deal
with acid rain were met with a rising crescendo of
skepticism and calls for nothing more substantive
than further study. A strange coincidence of
circumstances—world uncertainty over oil supplies, the holding of American hostages in Iran,
and the approaching presidential elections—
caused the controversy to emerge in the spotlight
of the congressional committee.

The catalyst that brought Eckhardt, Costle,
Coleman and dozens of others together was legis-

lation proposed by the administration of President Carter that would have resulted in a substantial increase in the burning of coal in those areas that were already subjected to high concentrations of acid precipitation. The Carter administration had decided to drastically cut the oil and natural gas consumption of the nation's utilities. Scattered throughout the United States, but concentrated mainly in the Ohio Valley and the Northeast, are dozens of electricity-generating power plants fueled by oil and gas. Many of these units are old, many are small, and many were converted from burning dirty, high-sulfur coal to less dirty, low-sulfur oil or gas in the early 1970s. In this way the utility companies were able to meet the new federal clean air regulations without having to install air pollution control equipment on their plants.

In early 1980, the United States was importing approximately 8 million barrels of oil a day, about 40 percent of its oil supply.[2] Coal, which represents nearly 90 percent of the nation's energy reserves, was being used to meet only 20 percent of its energy demand.[3]

A plan, therefore, that would force a number of plants to switch to burning coal and that would reduce the nation's oil consumption by about 1 million barrels a day seemed attractive to the administration. And it was especially attractive to an administration that was looking for something to offer as tangible evidence of the president's

Top 20 Coal-Fired Power Plants in the U.S.A. Ranked According to Total SO$_2$ Emissions in 1979

Rank	Plant	State	Estimated SO$_2$ Emission Thousands of Tonnes/Year
1	Paradise	Kentucky	372.5
2	Muskingum	Ohio	340.2
3	Gavin	Ohio	339.5
4	Cumberland	Tennessee	289.7
5	Monroe	Michigan	264.9
6	Clifty Creek	Indiana	263.7
7	Gibson	Indiana	261.1
8	Baldwin	Illinois	257.9
9	Labadie	Missouri	224.0
10	Kyger Creek	Ohio	205.5
11	Bowen	Georgia	202.6
12	Conesville	Ohio	186.8
13	Mitchell	West Virginia	186.2
14	Hatfields	Pennsylvania	167.3
15	New Madrid	Missouri	164.0
16	Sammis	Ohio	160.7
17	Wansley	Georgia	159.7
18	Homer City	Pennsylvania	159.1
19	Johnsonville	Tennessee	157.9
20	Gaston EC	Alabama	154.8

Total 4,518.1

Source: Province of Ontario, A Submission to the United States Environmental Protection Agency Opposing Relaxation of SO$_2$ Emission Limits in State Implementation Plans and Urging Enforcement, 12 March 1981; Expanded 27 March 1981, Ontario Ministry of the Environment, 1981, p. 17

effectiveness at a time when the Iranian militants were holding our embassy workers hostage, and a tough reelection campaign was getting under way.

Since little of Carter's ambitious "war" on the nation's energy crisis had made it through Congress, the administration hoped that passage of the coal conversion bill would create the impression that some progress was being made. In addition, the bill would result in the mining of more coal—especially coal close to the converting power plants. And since there were a good many miners out of work at the time, a pro-coal bill might help Carter to win some key states in the November election. Thus, the "oil backout bill"— Capitol Hill jargon for getting power plants out of the oil-burning business—was born.

The president's original plan would have required about fifty power plants operated by thirty-one utilities to convert to coal. It would have spent $3.6 billion in grants to the utilities to help them cover the cost of converting and would offer another $500 million to help the utilities buy pollution controls for the converted plants and to fund acid rain research. The utilities, however, would not be required to put controls on the converted plants.[4]

As described by its foes, the plan would have forced a collection of power plants in the heart of the country's acid rain belt to begin generating about 26 trillion watts of electricity—6 percent of the national total—by burning up to 50 million tons of coal a year without any pollution control requirements.[5] Although the proposal would be revised and somewhat limited in scope, the crucial question it posed remained the same: Is the goal of burning less oil worth the environmental risk of burning more coal in an area already suffering the effects of coal-caused pollution?

Characteristically, the Carter administration presented nothing close to a unified front when the coal conversion issue hearings were held by Eckhardt's House Subcommittee on Oversight and Investigations in late February. The Department of Energy, which had drafted legislation without consulting the Environmental Protection Agency, was eager to produce a bill that utility companies would support. An earlier version had been rejected by the utilities on the grounds that it would have cost them too much to convert their plants. In the version the Energy Department prepared for Congress, the money for conversion was substantially increased, and there were no requirements for pollution controls on the converted plants.

When Douglas Costle at the Environmental Protection Agency (EPA) finally saw the bill, he directed EPA staff members to draft a position paper that labeled it unacceptable. Costle and the EPA maintained that the proposal would cause acid-forming pollution to increase by 25 percent in New England, the area of the country already the most severely affected by acid rain. The EPA immediately proposed an alternative that would forbid any increase in pollution in the area of a plant converted to coal. Thus the battle lines were drawn.

In the first congressional hearings before Representative Eckhardt's Subcommittee on Oversight and Investigations, Costle tried to minimize the rift in the administration that had developed as a result of the environmental consequences of the oil backout bill. "There are differences," Costle testified. "I think that is probably about as far as I care to go on the subject this afternoon."[6]

Although Eckhardt responded that "I understand you may not want to wash your linen here before this subcommittee," he proceeded to have Costle and Lynn Coleman of the Energy Department do just that. As they did, they recreated in capsule form the debate over what the nation should do about acid rain.[7]

Eckhardt began: "It seems to me . . . when we are talking about expending a considerable sum of money to move from oil to coal, at a time when we are already threatening the environment with existing conditions, we had better look a long time and think very hard before we increase that."[8]

Costle testified in reply: "I do not think that we should be increasing emissions in New England at this time. I think we know enough now [about acid rain] to decide whether that is a good or a bad idea. At least for my part I have decided—I think it is a bad idea."[9]

In turn, Coleman argued that the option of converting power plants from oil to coal is one the nation must follow in view of its energy dependence on other countries. "One way or another we must take advantage of abundant coal supplies," Coleman said. Congress should not take actions to remedy acid rain now "because we don't know enough about it and we might pick the wrong solution right now."[10]

The two sides remained entrenched in other hearings. Later in the spring, before the Senate Subcommittee on Environmental Pollution chaired by Senator Edmund S. Muskie, Democrat of Maine, Costle and Senator Robert Stafford, Republican of Vermont, had this exchange:

STAFFORD: Basically, Mr. Costle. . . the technology exists or is now on the drawing board to remove the pollutants [that] cause acid rain. Am I correct in that assumption?

COSTLE: I think that is a correct statement.

STAFFORD: Are there any in the Carter administration who oppose the installation of such pollution control systems?

COSTLE: I think there is a natural tension between those people who are charged with producing more kilowatts and finding other ways to do it and finding other things to use than oil and those of us who are concerned about the environmental effects of doing it. I think there is a lot of pressure, obviously, from utilities to keep their costs as low as possible. That is really the area of debate. I don't think there is any debate about whether technology exists."[11]

During the months of hearings, the fight was joined by people outside the government on both sides of the issue. At the forefront were the Canadians, who were publicly angered by the prospect of Congress passing a law that deliberately increased their share of acid-forming pollutants.

Acid rain is a hotly debated topic in Canada because most of the population of the country lives close to areas that have been hit hard by acid and because many Canadians derive their livelihood from industries that could be hurt by acid precipitation. Tourism—much of it in the lake country vulnerable to acid rain—is worth $5 billion a year to Canada.[12] In addition, 10 percent of all Canadians depend on jobs in forest industries which are potential victims of acid rain's effects on tree growth.[13] The highly explosive issue has catalyzed the formation of a surprisingly broad-based political coalition of old-line environmentalists, commercial and sport fishermen, loggers, resort owners and users, scientists, utility officials, government officials, and just plain folks whose motto, "Stop Acid Rain," is also disseminated in French—*Halte aux Pluies Acides*. Many Canadians are angry about the pollution coming from the United States, and their rage is vented across the border.

"There will be an explosion of Canadian anger

the likes of which our neighbor has never seen if we have to wait fifteen years for abatement and thousands of lakes to die,"[14] said John Fraser, then Canadian minister of the environment, in a reaction to the proposed oil backout legislation.

Peter Towe, former Canadian ambassador to the United States, said in a May 1980 interview that "we understand the desire to reduce dependence on foreign oil, but the [Carter] administration's proposals are a step in the wrong direction from what we had agreed to in the past."[15]

Raymond Robinson of Environment Canada, the Canadian counterpart to the EPA, was equally outraged by the proposed oil backout bill. "The situation brings to mind the pledge made by the Prime Minister of Canada, President Carter and other leaders of western nations at the June 1979 Tokyo Summit Conference to increase their use of coal. That pledge included an important corollary—that increased use of coal would be achieved without damaging the environment," said Robinson.

"When we in Canada examine the conversion program . . . we do so in the sure knowledge that the environment of sensitive regions of our country has already been damaged by acidic precipitation. Possible [U.S.] increases in emissions could wipe out Canadian progress."

The conversion plan "presents a singular legislative opportunity to begin to deal with the [acid rain] situation," Robinson concluded, but "to put the matter bluntly, there does not appear to have been any meaningful attempt to consider the very real impacts . . . on Canada."[16]

In lobbying the State Department, the EPA, and members of Congress, Canadian officials demanded that any coal conversion bill require a reduction in pollution emissions in the United States. "The status quo is not enough," said George Rejhon, the environmental affairs official in the Canadian Embassy in Washington at the time. "We've gotten 50 percent reduction in emissions from Inco [a Sudbury, Ontario, metal smelting company], the continent's largest sulfur polluter. We would be very happy if you could say the same about Ohio."[17]

Other foes of the oil backout plan singled out the dismal pollution record of Ohio as one reason that the bill should be rejected. Robert Rauch said that if the power plants began burning coal, the additional sulfur dioxide and other pollutants discharged by the plants would increase pollution levels in Ohio and the New England states to the point where no new sources of pollution—a new factory, for instance—could be added without violating federal pollution laws. That stifling of economic growth would lead, he said, to a campaign to relax the pollution limits similar to an effort underway in Ohio where utilities were seeking relaxation of limits on twenty-one power plants. "When coal comes in, the pressure will increase to relax the emission limits and we'll see the 'Ohio-ization' of New England,"[18] Rauch predicted.

Richard Ayres of the Natural Resources Defense Council told Senator Muskie's Subcommittee on Environmental Pollution that the oil backout bill "contemplates a federal subsidy for degradation of the environment and public health. If there are to be subsidies, the Congress should use this opportunity to improve environmental quality. You are challenged, should you decide to legislate, to restore the balance to meet the American people's desire for an energy policy which does not sacrifice environmental quality."[19]

The *New York Times* editorialized in March 1980 that "Mr. Carter might wisely take some of the billions he proposes to lavish needlessly on conversion and buy, instead, an umbrella against acid rain."[20]

Friends of the conversion plan included, of course, the utilities. Samuel Huntington, general counsel for the New England Electric System, testified before the Senate subcommittee that conversion to coal burning would not cause significant increases in emissions. "We believe that the current controversy over acid rain and what the appropriate mechanism for dealing with acid rain is, should not be used to block a program which will have significant economic and national security benefits and a de minimis environmental impact."[21]

During the hearings, both Huntington and representatives of the Electric Power Research Institute presented elaborate analyses of the acid rain situation. They concluded that not only was no action warranted other than further study, but no one had produced conclusive evidence to demonstrate that an acid rain problem really existed.

James F. Smith, chairman of Orange and Rockland Utilities, a New York elecric utility, wrote in a 1980 letter to the *New York Times* that acid rain "is an environmental mystery. Should I therefore as the chief executive of an energy utility ask our 625,000 customers to await the answer to a mystery while they suffer the economic ravages inflicted on them by OPEC oil extortion? Are they—all of us—to stand by helplessly suffering for years while the government ponders the heavens?"[22]

While some witnesses focused on the central question of whether the claimed economic and national security benefits of the oil backout bill justified the resulting increase in acid precipitation, much of the discussion revolved around peripheral disputes. The Department of Energy and the Environmental Protection Agency agreed from the beginning that the coal conversion would inevitably cause an increase in acid-forming pollutants because most coal burned in power plants contains more sulfur per unit of energy than oil does. But they spent considerable time refuting each other's calculations of exactly how many hundreds of thousands of tons more sulfur dioxide and nitrogen oxides would be pumped into midwestern or northeastern skies each year. In other words, instead of debating whether or not to legislate an increase in pollution, most witnesses and committee members argued about how much of an increase would be acceptable.

A dubious highlight of the Eckhardt hearings in the House was a discussion of the likelihood that the increased pollution from the converted plants near the east coast would be blown harmlessly out to sea. In fact, a draft memorandum on the proposed bill prepared by the Energy Department states, "A large percentage of [the increased emissions] are likely to be dispersed over the ocean by prevailing winds and thus will not impact the region itself."[23]

When the Environmental Protection Agency heard about the draft memorandum, it prepared a paper entitled, "Emissions Increases Are Not All Blown Out To Sea." The EPA paper said the Energy Department had assumed conventional wisdom about the wind; in fact, most of the time the winds would deposit the pollution on New England, especially in the summer when the rate of conversion of sulfur dioxide to acid is believed to be most rapid. A third or less of the pollutants would be blown out to sea,[24] the EPA said. These documents were duly recorded and entered in the permanent record of the hearings, and there was a brief debate in which the participants tried to decide if one-third is a "large" percentage.

More disturbing, though, was the testimony about the involvement—or more precisely, the lack of involvement—of environmental protection officials in creating the administration's bill. During the Eckhardt hearings, Representative Albert Gore, Jr., Democrat of Tennessee, began his dissection of the Energy Department by pointing out that the department had made no mention of acid rain or increased pollution in the bill or the memorandum that accompanied it. Gore added that the EPA was not consulted in any way with the development of that conversion proposal. "That would seem to evidence a lack of coordination, a lack of awareness of this aspect of the problem,"[25] he told energy department counsel Lynn Coleman. In fact, congressional testimony revealed that the Department of Energy went out of its way to consult with private utilities in drafting the oil backout legislation, but let the EPA see the bill only the day before it was sent to the Office of Management and Budget for final approval on its way to Congress.

The Energy Department does have its own environmental experts, and Lynn Coleman, counsel for the agency, testified that they, at least, had been consulted in formulating the bill. In fact, they were not. A report from the staff of the Eckhardt subcommittee stated that the DOE environmental office "was not involved in writing or reviewing the legislative proposal."[26] Only later, after the department's environmental staff raised objections to the bill, were they consulted.

Congressman Gore was disturbed by these revelations. "You know, time and again we see an initiative brought up which just takes a single perspective and pushes it to the exclusion of all others," Gore mused. "Then we . . . hear from Canada and we hear from other people impacted by this. The question is, Why did you not think about this?"[27]

Coleman's explanation for the who-didn't-talk-to-whom problem was that the Energy Department wanted to draft a bill that would attract the least amount of opposition and become law in 1980. "At this time, the focus was on what was the most practical method of developing a coal conversion program," Coleman testified. "From my own point of view, I thought the issue on the environmental part of it had been settled for the purpose of this program."[28]

The issue, however, was not at all settled in the minds of the congressional committee members, and the environmental groups that were lobbying hard against the bill were optimistic that it would not be passed. Rafe Pomerance, at the time legislative coordinator for Friends of the Earth in Washington, said in an interview that while the public split between Carter and the EPA was unfortunate, Congressional sensitivity to public awareness of the acid rain issue would be most important in determining the bill's fate. "Bureaucrats contradict each other all the time. But the coal conversion bill just didn't deal with the acid rain issue, and so it's probably on death row this session."[29] In the end Pomerance was right, but for a while the outcome was uncertain.

In June 1980 the Senate overwhelmingly passed a scaled-down version of the oil backout bill that would subsidize the conversion of eighty power plant boilers to burn coal and would provide $600 million to help utilities install smokestack scrubbers and coal-washing or other sulfur-removing equipment. Robert Byrd, Democrat of West Virginia, then Senate majority leader and a stalwart supporter of coal mining interests, led the pro-conversion forces. The bipartisan opposition was led by Senator Paul Tsongas, Democrat of Massachusetts.

Tsongas wanted to spend federal money to help the utilities only if pollution from the converted coal-burning plants was kept at or below the oil-burning level. As an alternative, Tsongas suggested, utilities could be allowed to increase emissions from the converted plants as long as the increase was offset by decreased emissions at other nearby facilities. This so-called "bubble" approach would mean that while pollution from an individual source might increase, the total emissions in the region would not increase, and thus acid precipitation would not be aggravated.

Tsongas's alternative plan failed in a close vote

in the Senate Energy and Natural Resources Committee, which was deeply divided over the environmental issue posed by the bill. Tsongas tried again on the floor of the Senate. Republican Robert Stafford of Vermont supported him in the debate, arguing that tight state restrictions on air pollution in the Northeast had left room for future industrial growth without violating federally mandated air pollution limits. Stafford claimed that such growth is vital to the Frostbelt states in a time when people and investments are moving to the South and West. But the additional pollution from the converted plants would wipe out that breathing space, he said.

"We paid for our clean air with hard cash and hard work, but the federal government under this bill takes that clean air away from us," Stafford argued on the floor of the Senate. "All that seems to matter is that the god of energy prevails over all other rights and interests."[30]

Despite the efforts of Stafford and Tsongas, the Senate passed the bill by an eighty to seven vote.[31] In the House, however, coal conversion was buried in the Subcommittee on Energy and Power and was never resurrected. Some members objected to the more than $4 billion price tag, referring to the federal aid as "corporate welfare." Others could not accept the environmental degradation that would come with conversion.

The results, then, of this first test of American will on the emerging acid rain issue were inconclusive. The Senate bowed to expediency and the pressure of energy interests and decided that acid rain was not yet an issue that required serious consideration. The House, faced with the same opportunity to deal aggressively and creatively with the pressing but conflicting demands of energy independence and environmental quality, chose the disheartening option of doing nothing. The environment was spared another insult by default, but the nation's leaders failed to make things any better.

7.   *A Lethal Legacy*

It's a ramshackle building, set down in the woods at the edge of a stream. From the two-lane road on the outskirts of Ithaca, New York, it doesn't look like much, but inside a group of scientists is tinkering with evolution. They are looking for a trout that can live in acid.

The scientists place normal baby trout into tanks of water and add a measured amount of acid. Though nearly all of the fish become deformed and die, a few survive, reach adulthood, and lay eggs. This second generation of fish is exposed to a stronger concentration of acid. Again, almost all die, but a few somehow withstand the effects of the acid and live to reproduce. If the process is carried on through a number of generations, with exposure to ever higher concentrations of acid, the scientists believe that a genetically new fish may emerge—a mutant trout that will be able to survive in the tens of thousands of lakes and streams in North America and the rest of the world that are being poisoned by acid rain.

So far Carl Schofield, the Cornell University biologist, and his research colleagues have not been able to find the acid-resistant trout that can live in lakes as acidic as those in the Adirondacks. The fact that they are trying, however, is testimony to the belief among many scientists, government officials, and environmentalists that acid rain and its damage will be a part of our world for a long time to come.

There is a consensus of pessimism that even in the unlikely event of a move by government and industry to aggressively clamp down on the discharge of the pollutants that form acid rain, the problem will continue to grow worse in the remainder of this century.

"We are losing the war against acid rain. We are not even standing still," says Robert Rauch of the Environmental Defense Fund. "Each day allowable sulfur dioxide emissions are increasing in this country. Each day the situation is deteriorating. The gap between rhetoric and reality is opening into a chasm."[1]

The evidence for Rauch's lament is clear. After decreasing during the first half of the 1970s, emissions of sulfur dioxide began increasing again during the second half of the decade. Emissions of nitrogen oxides, the other main precursor of acid rain, also continued their decade-long upward trend.

New coal-fired power plants going on line are required to remove from 70 to 90 percent of the sulfur from their smokestack gases. But more than 90 percent of the coal plants in operation in the United States have no controls and in the past two years, there has been an accelerating trend on the part of federal pollution officials to allow these old, polluting plants to burn coal with even higher sulfur content than they are currently burning. Sooner or later, these pollutants and their acid descendants will enter the aquatic ecosystems of the United States and Canada and continue the degradation already under way.

"We have lost lakes and undoubtedly we will lose more,"[2] says Ellis Cowling, the North Carolina State University scientist and head of the National Atmospheric Deposition Program, a private organization that analyzes samples of acid rain, snow, and dry fallout from thirty-two U.S. states.

The idea of developing an acid-resistant trout as a solution to the acid rain problem may make as much sense as making bulletproof vests a substitute for law enforcement. But it is the kind of answer, unfortunately, that gains credence in a time when national policy is shifting away from an ethic of maximum environmental protection and toward an ethic of economic preservation that takes some ecological destruction for granted.

It is also an answer, however, that evades the real questions posed by acid rain. Acid rain is a problem of gargantuan dimensions because it is the direct result of energy production and motorized transportation—two of the most vital activities of an industrialized nation. Attacking the causes of acid rain would mean making difficult political and social decisions about the way these activities are carried out.

Different groups have proposed a variety of solutions to the problem that we can think of, for a moment, as a leaking basement faucet. We hear those who see water on the basement floor and say the solution is to wait until the first floor is flooded to make sure that the problem is real. Carl Bagge, president of the National Coal Association, says: "We are concerned with acid rain . . . but at this point it is just a concern. We don't have any hard data to make a connection to coal. I hope the concerns are not conceived by the public as problems until they are identified as such."[3]

We hear those who say the answer is to let the faucet run but pump out the flooded basement. James F. McAvoy, formerly chief environmental protection official for the state of Ohio—the state with the worst sulfur dioxide pollution record in the nation—seriously suggests, for example, that pouring lime in lakes damaged by acid rain "is pragmatic and may be a more cost-effective solution than sulfur removal at the source."[4]

We hear those who see the leak but claim that their faucet is not leaking. The officials of such heavy coal-burning states as Ohio, Michigan, West Virginia, and Pennsylvania decry the acid rain falling on their territory and blame it on their next-door neighbors or nature.

And we hear those who say the way to stop the flood is to catch the water running from the faucet in a bucket and dump it elsewhere. Some people in this group, such as those who want to install equipment to remove sulfur from power plant smoke or from the coal itself, opt for a big bucket. Others, such as former U.S. Energy Secretary James Schlesinger, suggest using no more than a child's pail. Schlesinger's bold idea for stemming the spread of acid rain was to control sulfur emissions from a power plant only when the wind blows in the wrong direction—an idea that Senator Edmund Muskie of Maine described as the "rhythm method of controlling air pollution and about as effective."[5]

The voices that are fewest and least attended to today are those who take a fundamentally different approach to the problem. In order to stop the flood, they say, turn off the faucet; in other words, stop acid rain at its source. Stop burning so much coal and gasoline, reduce demand for energy, and find alternative means of supplying the energy that is really needed.

"If we are bent on increasing energy consumption and doing it with coal," acid rain researcher Eville Gorham told a Congressional subcommittee, "then I suspect we will pay a terrible price for it sooner or later and within the lifetime of my children. Those of us who have lived over fifty years have seen a marked rise in our living standards and our energy consumption. I have not been impressed with a similar rise in our personal happiness."[6]

So far the acid rain debate has been dominated by the people who advocate using buckets—those who advocate burning coal, even much more of it, and capturing the sulfur and nitrogen pollutants before they contaminate the atmosphere. A number of processes are available for preventing as much as 90 percent of the sulfur that starts out in a lump of mined coal from escaping out the top of a smokestack. Sulfur can be removed from the coal as it leaves the mine, as it burns in the power plant furnace, or as its smoke enters the stack. Nearly all the existing sulfur pollution control in the world takes place just before the smoke leaves the plant.

The process known as "wet scrubbing," which is the oldest and most commonly used means of removing sulfur dioxide from power plant smoke, takes place inside a massive structure about six hundred feet long and ten stories high. Inside is a system in which the hot gases from the coal fire are directed over a vacuum cleaner through which a shower of water containing fine particles of lime-stone is passed. This alkaline spray converts the sulfur dioxide in the smoke into a semisolid sludge about the consistency of toothpaste that is difficult to handle and environmentally hazardous. From 1 to 3 percent of the energy produced in burning 10,000 tons of coal a day in a typical 900 megawatt power plant is siphoned off to power the scrubber.

The Operation of a Coal Scrubber

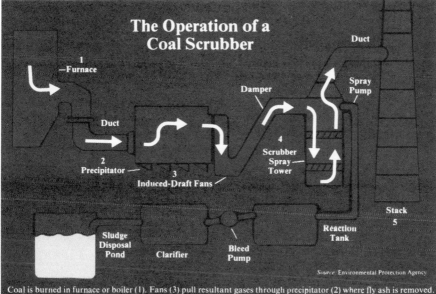

Source: Environmental Protection Agency

Coal is burned in furnace or boiler (1). Fans (3) pull resultant gases through precipitator (2) where fly ash is removed. Damper directs gases to scrubber spray tower (4) where slurry of water and chemicals is sprayed to remove SO_2 and remaining ash. Clean gases then go up stack (5). Liquid chemical used to absorb SO_2 drains into reaction tank where sulphur is removed through a chemical process. Bleed pump routes it to clarifier from which it drains to sludge disposal pond.

Other forms of the scrubbing process have been developed in recent years. "Dry scrubbing" uses much less water and energy and is less expensive than wet scrubbing. Still another process, known as regenerative scrubbing, avoids the problem of great volumes of sludge by converting sulfur dioxide to pure sulfur. Some researchers have noted that this sulfur could be collected from midwestern power plants and shipped to sulfur-poor regions of North America. In fact, Pennsylvania highway crews are already using waste sulfur to replace part of the asphalt needed for road resurfacing. The Canadians are intensively researching ways to market sulfur waste by-products from power plants and smelters and have made such research a priority in their efforts to find comprehensive ways of dealing with the acid rain problem. Unfortunately, neither dry scrubbing nor regenerative scrubbing are being used in a full-size U.S. power plant.

Only about 15 percent of the nation's coal-fired energy comes from power plants equipped with sulfur dioxide scrubbers.[7] The main reason for the sparse use of scrubbers is their cost—as much as 15 percent of the cost of a new plant, which was about $800 million in 1980.[8] A typical new coal-fired plant equipped with scrubbers emits more than 33 million pounds of sulfur dioxide and nearly 28 million pounds of nitrogen oxides each year. Installing scrubbers in an old plant is usually more expensive than scrubbers in a new one and must be financed over a shorter period of time; consequently, the number of old plants with scrubbers is miniscule.

Until the U.S. Clean Air Act was amended in 1979 to include "New Source Performance Standards" that require removal of from 70 to 90 percent of the sulfur burned in a new plant, utilities usually chose to burn low-sulfur western coal in order to meet legal sulfur dioxide limits. Ironically, the pollution control requirement of the amended law may work against reducing the acid load in the atmosphere. Utilities have found they will benefit by keeping older, polluting plants in use past their planned retirement dates rather than making the huge investment needed to build a cleaner, new plant.

Pollution control of nitrogen oxides is far behind sulfur dioxide control technology. Although some new processes are being investigated, none has been tried on a full-size power plant. The lack of effective nitrogen emission controls is becoming an increasingly serious problem because new power plants operate at much higher boiler temperatures than older units and thus produce more nitrogen oxides. Power plants built since 1970 have accounted for most of the alarming increase in nitrogen oxide emissions in the United States during the last decade. Nitric acid is expected to make up an ever larger proportion of acid rain as more new plants are built to replace the old.

One developing technology offers the hope of

controlling acid-forming sulfur dioxide and nitrogen oxides at the same time. Called "fluidized bed combustion," this process changes the way the coal is burned in the power plant furnace to prevent sulfur dioxide and nitrogen oxides from forming. Babcock and Wilcox, an energy engineering company more widely known for having built the ill-fated nuclear reactors at Three Mile Island, is one firm involved in developing the fluidized bed process.

Crushed limestone is held in swirling suspension in the coal furnace by air blasted in through thousands of tiny holes in the floor. When crushed coal is injected into the furnace, it burns on top of and mixes with the crushed limestone. The limestone absorbs the sulfur from the coal, producing calcium sulfate and calcium oxide, hard, dry, grainy by-products that are removed continuously from the boiler as new limestone is added. Ash that flies out the top of the furnace is captured in a baghouse—a giant vacuum cleaner—and little more than hot air is actually released from the smokestack.

The temperature of the furnace in the fluidized bed process is low enough to prevent the formation of most nitrogen oxides. If engineers can raise the efficiency of the process to that of conventional power plant boilers, they believe that it will be economically viable. A fluidized bed coal plant has been operating for more than a year on the campus of Georgetown University in Washington, D.C.

Another fluidized bed project, sponsored by the state of Ohio, ran for only sixty seconds before it broke down indefinitely.[9]

By far the most environmentally and economically attractive technological option available for reducing sulfur dioxide emissions is known, appropriately, as coal washing. Sulfur can be separated from coal that is crushed and subjected to a number of purification procedures that use gravity to remove the sulfur and other impurities. Coal washing, which is now done only at a few power plant sites, removes from 25 to 40 percent of the sulfur from high sulfur coal and produces a lighter coal with substantially less ash content.

The Electric Power Research Institute has concluded that coal washing may pay for itself. If it is done at the mine, the cleaned, lighter coal is less expensive to ship. It produces less ash when it burns, which results in less boiler maintenance expense and less money spent burying waste ash. Ohio may become the first state in the nation to require washing of all coal burned within its borders, and some environmentalists have proposed that the federal government require coal washing at every power plant in the acid plagued East and Midwest. The Bruce Mansfield power plant in Shippingport, Pennsylvania, which is one of the few plants in the country using wet scrubbing and washed coal, has emissions that are 92.1 percent sulfur free.[10] Though chemical processes exist for removing even more sulfur, they are still expensive and have not yet proven their economical viability on a commercial scale.

Perhaps the most innovative method for ridding coal of sulfur was the brainchild of Henry Tsuchiya, a microbiologist in the chemical engineering department of the University of Minnesota. He and colleagues in California have found a possible way to use bacteria to digest sulfur from coal. *Thiobacillus ferrooxidans*, an ordinary looking, rod-shaped bacterium, was discovered in acidic mine drainage several decades ago. It leaches sulfur from coal and copper and converts it to sulfuric acid, which can be washed away from the coal. Tsuchiya theorized that one possible solution to the acid rain problem would be to use *Thiobacillus ferrooxidans* to take sulfur out of coal before it is burned.

In a study financed by the U.S. Department of Energy, Tsuchiya and his colleagues found that the bacteria could remove 90 to 98 percent of the sulfur from a slurry of coal and water in eight to twelve days. Then the funding for the research project was stopped. "The people in Washington just don't feel it is worth it," says Tsuchiya. "Money's tight. I suppose eventually they'll pick up on it again—it's too good an idea. Eventually they'll realize that the precombustion removal of sulfur is the only way to go."[11]

Those who advocate the removal of acid-forming pollutants before they are released into the atmosphere may be trading one set of problems

for another. Continuing to burn coal and depending on technology to reduce the magnitude of the
resulting acid precipitation carries a hefty environmental and economic price tag. David Brower, a
well-known conservationist, maintains that rely-
ing on technology alone to solve the acid rain
problem is misguided. "We should not take science
and its child, technology, as savior. In science,
there is no moral context. What we need is new
human will, not new careless technology."[12]

Waste limestone sludge from scrubbers poses
environmentally hazardous and potentially expensive disposal problems. Neither coal washing,
fluidized bed combustion, nor scrubbing can prevent other dangerous contaminants—metals, radio-
active particles, and carcinogens—from leaving
the smokestack. In addition, no process exists to
control the emission of carbon dioxide, an inevita-
ble by-product of coal burning that is emerging as a
new global pollution threat. And the net cost of an
aggressive program of coal washing, new combustion technology, and scrubbing that would cut
U.S. sulfur dioxide emissions in half has been
estimated at about $4 billion a year.

Despite legislative proposals by lawmakers
such as U.S. senators Daniel Patrick Moynihan of
New York and George J. Mitchell of Maine to
mandate reductions in acid-forming emissions,
there are no signs that either the federal government or the power industry is prepared to support
that kind of commitment. In fact, the utility

industry and its supporters have stated that controlling acid precipitation by reducing emissions is ill advised.

Ohio Governor James A. Rhodes dismissed the critics of emissions from American power plants as "no-growth environmentalists" and accused them of having "latched on to acid rain as a rallying cry for a new wave of environmental hysteria."[13] James McAvoy, former head of the Ohio Environmental Protection Agency, testified about acid rain in 1980 before the House Subcommittee on Oversight and Investigations. Only one of Ohio's more than thirty power plants has scrubbers, and the state pumps about 3 million metric tons of sulfur dioxide into the air each year—approximately 10 percent of the nationwide total.[14] McAvoy defended the utilities and advocated even more coal burning in the future. Here are excerpts from his remarks to the congressional subcommittee.

● "Despite the reported effects of acid rain on the environment we cannot afford to overreact to preliminary data, especially in light of our grave energy needs today....

● "There are reports that nature may account for up to two-thirds of the total acid-producing air contaminants. Nature contributes to the acid-producing compounds in the atmosphere from both biological processes on land as well as chemical processes on the sea in the form of windblown sea salts. Few people realize that lightning flashes in thunderstorms can create nitric acid....

● "Right now, however, man-made sulfur dioxide emissions are getting most of the blame for the acid rain problem....

● "In behalf of our Governor, we say we reject the hypothesis that while [study of acid rain] is going on there should be any moratorium whatsoever on increased combustion of coal. We feel we must go in the opposite direction."[15]

In an exchange with Republican Marc L. Marks of Pennsylvania during the same subcommittee hearing, McAvoy refused to acknowledge that acid rain is a serious environmental threat. Marks referred to a Council of Environmental Quality report calling acid rain one of the foremost global environmental problems and said to McAvoy, "You don't mean to say you or the people in Ohio disagree with that?"

MCAVOY: "My remarks are not in conflict. I said repeatedly throughout my testimony we agree there may be a problem. What we are saying is that we do not . . ."

MARKS: "You say there may be a problem. Aren't you willing to suggest there is in fact a problem and that your responsibility and the responsibility of the state of Ohio is to try to resolve that problem? . . . You will admit it is a very serious problem?"

MCAVOY: "No, sir."

MARKS: "You won't?"

MCAVOY: "No, sir."[16]

In 1981 McAvoy was nominated by Ronald Reagan to head the Council on Environmental Quality, the "environmental conscience" of the executive branch, and he became the administration's top acid rain theorist. Environmental groups and members of Congress protested so strongly against the nomination that Reagan withdrew it. McAvoy, however, still has a job on the council staff.

Turning out in force at congressional hearings and taking their case to the media, the coal-burning utilities have maintained that the United States should not act to reduce sulfur dioxide and nitrogen oxide emissions because there is no unassailable proof that acid rain is causing ecological damage or that acid rain in a given place is caused by power plant emissions that originate elsewhere. "Reasonableness demands that we take the necessary time, make the necessary effort, and spend the necessary money to find the answers before we inflict the additional significant costs of control strategy on the public,"[17] says William N. Poundstone, executive vice-president of Consolidation Coal Company. And even if at some time in the future acid rain is found to be an environmental problem that cannot be ignored, Poundstone says, we should take the time to do "a comprehensive cost-benefit study of potential control strategies" before acting.[18] Until then, he says, wait and study.

George Hendrey of the Brookhaven National Laboratory is more than a little tired of the ceaseless calls for more study. "The utility industry cannot accept the idea that they are causing acid rain and environmental damage," says Hendrey. "It doesn't matter what we demonstrate. They scream and claim they're not responsible. As long as research is going on, the final word is not in. It puts off the time for a determination of the amount of sulfur dioxide removal that is necessary.

"I don't know if we will ever know how much pollution is too much from any one particular plant. What matters is the total load to the atmosphere of these sources and the cost of the damage they cause,"[19] Hendrey concludes.

Ellis Cowling is another acid rain researcher who believes that we cannot afford to wait until serious damage to the environment has been proven. Instead, he believes, we should act immediately to stop the damage now occurring and to prevent future damage, even though we may not completely understand the mechanics of destruction.

"It appears to be a part of our national character to be powerful in our influence on the environment," Cowling says. "Modesty among Americans may be an unfamiliar trait but it may be a most becoming trait in the long run....I think we must undertake a reexamination of our energy and environmental policies. We are a nation which has exerted itself in the world in a substantial way. We are heavy consumers of energy and other natural resources. I think we need to become more modest in our demands upon the environment. The technology exists with which we can enjoy a very comfortable and quality existence without contaminating our environment so thoroughly."[20]

Robert Slater, former head of the Canadian Water Quality Board, makes the same points more bluntly: "To delay taking action on this matter until all the evidence is in . . . will simply result in our selling tickets to an autopsy."[21]

The array of arguments that support Slater's urgent tone reflect such wide-ranging concerns as human and ecological health, economic well being, and what can best be described as moral responsibility. The human health issue was raised again in 1981 by the researchers who discovered that the airborne sulfate particles which are found in varying concentrations in the air aggravate heart and lung diseases of Ohio River Valley residents. For much of the 1970s, investigators searching for links between air pollution and respiratory disease studied sulfur dioxide gas. In recent years, however, the focus of research has shifted to sulfates, and the 1981 Ohio River Valley study provided evidence that acid sulfate mist does indeed threaten human health.

The economic argument for concerted action now to reduce acid-forming emissions in North America is the one most frequently and forcefully presented by those who advocate immediate acid

rain remedies. Simply stated, this argument con-
tends that ignoring acid rain will result in more
cost to taxpayers than taking action to stop it now.
Recent estimates of the damage caused by acid rain
in the United States have placed the annual cost at
$5 billion for the corrosion of buildings, auto-
mobiles, artworks, and other human-made objects.
The damage to aquatic ecosystems, forests, and
farmlands has not been estimated, but it, too,
could amount to billions of dollars each year.

One of the lessons of the acid rain controversy
has been that environmental battles will have to be
fought primarily on economic grounds rather than
on the principle of preserving environmental
integrity. "The damage from acid rain to crops,
pasture, forests, recreation, buildings, posses-
sions—these are very real and very large costs,"
says acid rain researcher George Hendrey. "We
should add up all these costs and put a dollar value
on them. We have to convince the consumer that a
percent is taken out of his paycheck every time he
buys a commodity that is subject to acid rain—
every time his plastic and aluminum lawnchair
disintegrates or his storm windows won't slide
anymore because they are so corroded. We are
already paying for acid rain in an insidious way."[22]

Representative Andrew Maguire of New Jersey
believes that the prospect of economic damage will
be the decisive argument in persuading Congress
to act. "I would hope people would be sensitive to
the desirability of having lakes that are productive

and enjoyable and healthy," says Maguire, "but sometimes to win the argument we have to be able to say that in addition to those desirable things, it will cost us this many hundreds of millions or billions of dollars over this period of time if we in fact do not protect these resources.

"If we say this is what it will cost if we continue to make decisions which are not going to protect our resources, then a lot more people, perhaps a majority in the legislatures of these states or in the Congress, will take the actions required."[23]

In addition to the economic consequences of acid rain, acidification poses a long-term threat to the world's ecology. Each time a life form is eliminated, the genetic ability of nature to adapt to future assaults of climate, disease, or other catastrophic changes is diminished. The next species that is destroyed by acidification, for example, could be one that holds great potential benefit for humankind, such as the wild teosinte plant. A rare plant found only in a limited area of Mexico, teosinte was recently crossbred with domestic corn to produce a revolutionary new species of corn that is adaptable to almost all climates and resistant to almost all diseases.

"Any species we lose might have unsuspected value," says ecologist Eville Gorham. "If you knock off certain fish populations with acid rain, you might find one day that you need them to breed into another species you are managing for food or some other purpose.

"The tiny fruit flies that swarm around garbage cans—you might say let's get rid of them. It just so happens that those fruit flies are the major basis for most genetic studies. Studying fruit flies has taught us much of what we know about human inheritance."[24]

Gene Likens once described the ultimate cost, in terms of human survival, of the kind of ecological damage caused by acid rain. "One has to be very concerned about this kind of environmental insult on natural systems. There is a limit to the stress they can withstand. The forests and the lands are life-support systems. Without [them] to cleanse the air and water, to provide food for us to eat, our health is just as much in jeopardy as if something is affecting us directly."[25]

George Hendrey suggests that there is an argument for doing everything possible to stop acid precipitation that goes beyond dollars or self-preservation and stems from our relationship to the world in which we live. "When we decided that we didn't want our lakes to turn pea green with algae because of phosphate detergents, we didn't do it because of economics," says Hendrey. "We are paying for acid rain just by knowing that fish are disappearing, that when you get to a lake you'll find it dead with acid. It is another insult to our whole environment, like the disappearance of mammals in Africa or the dwindling of the whales or the contamination of groundwater by chemical wastes from industry. All of these things have to be considered together. It is the same as not knowing what a clear sky is anymore. In the United States we have contaminated the air to the point where we have to go to Greenland or South America to see a clear sky."[26]

The arguments for controlling acid-forming emissions are clear and compelling. However, the power that accepts these arguments and can bring pressure to bear on those who are capable of leading the industrialized nations through the perplexing difficulties in controlling trans-border pollution has not yet emerged.

Where is that power? We cannot look to Europe. Sweden and Norway have argued to no avail for longer than we have known of acid rain that they should not be victimized by the pollutants of England, Germany, and other European nations. Russia, in its silence and emphasis on industrial development, is an unlikely candidate for environmental leadership.

Canada, at least, sees the threat and has the will. In 1981 the Canadian government embarked on an aggressive campaign to persuade anyone who would listen that acid rain is a clear and present danger. The Canadians decided to take their campaign directly to the American people. They orchestrated press tours of acid-affected areas of Canada in a way that produced maximum coverage of the issue in Sunday newspapers throughout the United States. They sent a lobbyist to Washington and set aside $1 million to press their case—more than they had ever spent at one time to influence the policy of foreign nations. They sent speakers to American cities, distributed articles about acid rain, and handed out red "Stop Acid Rain" buttons at the border.[27]

In 1981 the Canadian House of Commons produced a thirty-eight point plan that warned of increasing acid-forming pollution in North America as a result of Canadian and U.S. industrial activity and called for aggressive control of the pollutants at their source. The plan recommended reducing emissions in all sectors of the Canadian and U.S. economies, intensifying research on acid precipitation, increasing public awareness in both nations, and expanding markets for sulfur and sulfuric acid by-products of pollution control.[28] Sulfur dioxide emissions from the Inco Ltd. smelter in Sudbury, Ontario, have been reduced from an average of 7,000 tons per day to just 2,500 tons.

While Canada may have the will, however, it lacks the clout. That leaves the United States as the only power in a position to assume the role of leadership on the acid rain issue. Is the U.S. government ready to acknowledge the threat and take the hard steps to control acid-forming emissions? By all indications the answer is no. Unfortunately, while the United States has the clout, it lacks the will.

The Canadians rightly fear that the Reagan administration intends to drag its feet while going

through the motions of doing something about
acid rain. U.S. Representative Clarence Brown,
Republican of Ohio, has even implied that Canada
may be trying to weaken the U.S. power industry
so that it can sell more Canadian power to
Americans. All the Canadian anger about acid rain
may come from "an exaggerated xenophobia...that
I find disturbing,"[29] he said.

The Canadian concerns are further justified by
statements such as the one made by Kathleen
Bennet, assistant administrator of the U.S. Environmental Protection Agency. Bennet told the
House Health and Environment Subcommittee
that the administration, which intends to spend
$18 million in fiscal year 1982 on acid rain
research, believes that more research is needed to
assess the problem. "The American people have
the right to expect that their government will not
impose an additional multi-billion dollar program
without first determining with some assurance
that the intended environmental benefits will be
achieved,"[30] said Bennet.

This Reagan administration policy continued
the official Washington line of procrastination
that had been established by the Carter administration. In a 1979 address to Congress on the
environment, President Carter detailed a number
of facts and suspicions about the damaging effects
of acid precipitation and then announced a
modestly funded ten-year research program "to
improve our understanding of acid rain."[31]

Critics were quick to point out the weaknesses in Carter's program, which authorized $10 million in research funding during the first year. The *Washington Post* editorialized: "In his recent environmental message to Congress, the president announced a new federal effort to assess the magnitude of this problem—but the announced duration is long and funding low. Since this air pollution doesn't respect national boundaries—as Canada's recent demand for an air pollution treaty attests—a much more intense international effort is clearly called for."[32]

The Acid Rain Assessment Program was established, but at the same time Carter was supporting the proposed synthetic fuels program—a program that threatened to cause a rapid increase in emissions of acid-forming pollutants in the northern plains. He also allowed Department of Energy officials to ride roughshod over the Environmental Protection Agency in the effort to force oil-burning utilities to convert to coal without requiring tight sulfur pollution controls.

Against this background of ambivalence, the U.S. Commission on Air Quality was making its own assessment of what should be done about acid rain. The commission, established by Congress in 1977, is charged with analyzing environmental, technological, scientific, and social issues relating to air quality. In findings published in January 1981, the commission's investigators stated that while "the extent to which sources such as power plants contribute to the [acid rain problem] cannot be positively determined, any reduction in precursor pollutants would lessen acid deposition.

"The United States has a wide range of options in planning its energy future, and decisions made now about energy policy may affect air quality significantly for the foreseeable future. The U.S. has the technology necessary to minimize energy-related air pollution and to do so effectively. Technology for controlling pollution from many processes has evolved to the point where it is appreciably more effective and less expensive than it was only a short time ago. This is especially true for controls available for the direct combustion of coal."

The commission found that the nation could burn more coal and still not increase the discharge of acid-forming pollutants by requiring the use of available pollution controls on new power plants. The status quo, however, would mean living with the level of acid that is now having a severe impact on aquatic ecosystems. To minimize acid rain, said the commission, the nation would have to use energy more wisely and efficiently.

"The commission concludes that the most effective means of controlling energy related air pollution is to conserve energy and to increase the efficiency with which it is used. Increased energy efficiency would substantially reduce U.S. dependence on imported oil and result in significant economic benefits. Conservation and improved energy efficiency can result in decreased emissions without affecting economic growth."

So there it was: The germ of a solution, the cornerstone of a sane policy, the key to ending the acid rain crisis. Yes, we do need more information about the causes and effects of acid precipitation, but we already have enough to make some reasoned decisions of unquestionable benefit to us, our children, and our world.

The first step is to adopt a new energy ethic. Energy prodution is responsible for 85 percent of the sulfur dioxide and 96 percent of the nitrogen oxides in the atmosphere over the United States. If we want fewer of these contaminants in the air, the place to begin is by producing less energy and using what we do produce more efficiently. "Aim now at the per-capita level energy use, roughly half of our own, of such backward places as Britain, Sweden, West Germany, Switzerland, and New York City,"[33] says David Brower.

The second step in the evolving acid rain control policy is to ask, once we know how much energy we really need, whether a new power plant in central Minnesota, northern California, or southern Ohio is the way to produce that energy. Environmentalist Robert Rauch answers that all proposed new plants should be subjected to an "alternatives analysis." Do we have to burn coal to solve our energy problem, for instance, or can we find some other, less damaging means? In California the Environmental Defense Fund found

that of the ten new coal-fired power plants
proposed by a utility, nine could be replaced with a
combination of cogeneration (using waste heat
from power plants), some solar energy instal-
lations, and conservation, which could be en-
couraged by interest-free insulation loans from the
utility.[34] "We are not talking about having people
turn down their thermostats or make sacrifices,"
Rauch told the House Subcommittee on Oversight
and Investigations. "That was not the point. The
point was whether we could find other techniques
to produce the same amount of electricity but
more cheaply and with less environmental consequences."[35]

The third step, once we have decided that we
have to burn some coal to make electricity, is to
require that coal-burning power plants be
equipped with the most efficient combustion and
pollution control technology available. Those who
believe pollution controls are too costly should
take a close look at the Japanese, who reduced
sulfur dioxide pollution by 50 percent from 1970
to 1975 while at the same time expanding their
economy and use of energy.[36] Furthermore, we
must look at the pollution that we reluctantly discharge into the air on a regional instead of a plant
by plant basis. We need, as Ellis Cowling says, to
start thinking of the atmosphere as a flowing river
in which pollutants entering at one point may be
carried downstream and emerge at another point
far from where they were first released. However

the lines are drawn—Midwest, Far West, North-east, Southeast, High Plains—and by whatever legislative or regulatory means necessary, Congress (or the House of Commons) and the government agencies responsible for protecting the air we breathe must recognize the relationship between smokestacks and prevailing weather patterns and place limits on the amount of acid-forming pollution that each region may produce.

Here, again, are the three elements of a successful acid rain policy:

● Use less energy and use it more efficiently.

● Burn only enough coal to supply the energy we cannot generate more easily and safely in other ways.

● Restrict the load of pollution that can be discharged into the air in any given region.

We are not locked in a coal/nuclear box, a choice between abandoning coal burning and its problems for nuclear energy and its problems. That is a false dilemma. Our survival does not depend on selling our future to either technology. With reasoned care and a blend of energy fuels and strategies, we can prosper economically and enjoy the tangible, essential benefit of an intact environment.

If the solution is so evident, why is there a consensus of pessimism among those who follow the acid rain controversy and among environmentalists. What leads David Johnson, a young lover of the mountains and an acid rain specialist for the Adirondack Park Agency, to lament the demise of a wilderness that was to be "forever wild": "I constantly feel the frustration that no way in hell is it going to get any better. I think all of us in the field are realists,"[37] says Johnson.

The answer is that those in the position to make the United States the power that leads the rest of the world toward a solution of the acid rain crisis have chosen not to follow a new ethic and embrace new priorities. Instead, they have chosen to revive the old ethic and the old priorities that resulted in the environmental dilemma we face today.

Ronald Reagan rose to the U.S. presidency preaching, verbatim, the formula for economic progress and social order in vogue thirty years before—precisely the time that the acidity of the world's rainfall began its rapid rise. Government is not in the business of telling business how to operate, he declared. Growth, expansion, using more energy and more resources to become powerful—these are the nation's priorities, not environmental integrity. Trees, explained Reagan in his campaign, are the major cause of nitrogen oxide pollution.

In this world view, the United States has no need for a President's Council on Environmental Quality, and within a month of his inauguration, Reagan was considering the elimination of the body that is responsible for developing ways of dealing with acid rain. Eventually he relented and nominated James McAvoy to run it.

In this world view, what America needed was James Watt, a foe of wilderness preservation, as Secretary of the Interior; Anne Gorsuch, a foe of tough pollution control, as director of the Environmental Protection Agency; and Secretary of Energy James B. Edwards, who offered this opinion about the nature of the acid rain problem: "I have a hard time buying the theory that acid rain is killing our lakes. All rain is acid. When you say 'acid rain,' it's a horrible sounding thing. We burned more coal in the twenties or thirties...and it didn't kill off lakes."[38]

To recommend funding levels for all government agencies, Reagan chose David Stockman, who told a meeting of businessmen in 1980: "How much are the fish worth in the 170 lakes that account for 4 percent of the lake area of New York? And does it make sense to spend billions of dollars controlling emissions from sources in Ohio and elsewhere if you're talking about very marginal volume of dollar value, either in recreational or commercial terms?"[39]

Given an administration that views acid rain in terms of a "very marginal volume of dollar value," the consensus of pessimism among those who are seeking a solution to the crisis is understandable. Sadly, the attitude that consuming energy and resources is more important than maintaining a healthy environment is not a new one. Earl Murphy, a law professor at Ohio State University and the author of two books on energy, environment, and public policy, tells a familiar story but one that has its roots nearly a century ago.

In the early 1970s, Murphy said, a Cleveland official who was worried about the condition of the city's shade trees discovered an old city document dealing with the same subject. The 1891 report blamed coal burned in factories and power plants for the deplorable condition of the trees. Noting that coal burning produced great quantities of oxides of sulphur, it stated that "sulfurous acid forms when sulfur oxides produced during combustion combine with water vapor in the air." The report also noted that the problem seemed to be regional because similar conditions had been found in Pittsburgh and Saint Louis.[40]

The 1891 document recommended three possible remedies: taller smokestacks, "smoke washing" devices, and "the use of a binding agent such as lime for trapping the pollutants." The reaction of the coal burning industries at the time, however, was that no "economically reasonable" or "technically feasible" method existed for cleaning up Cleveland's air.[41] City officials finally concluded that the best course of action—much like the one Carl Schofield is pursuing today—was to find and plant pollution-resistant trees.

"It is as if there is no new knowledge about man's responsibility to nature," said Murphy. "All that has been needed to be known was known long ago and . . . man has simply chosen not to act upon the basis of that knowledge. Nature is exploited;

the living and renewing environments are used as waste sinks; ecosystems are broken; and all of this is done with the knowledge of what is happening."[42]

Today there are signs that citizens are beginning to tire of the obfuscation and resistance of the Reagan administration and the coal and utility companies. Both on the state and the federal government levels, legislators are beginning to take the hard steps to control acid-forming emissions.

In Minnesota, where public awareness of the acid rain issue is greater than in many other states, concerned citizens have prompted legislators to act. For several years Minnesota legislators have warned of the impending acidification of the more than two thousand lakes in the vast and economically valuable Boundary Waters wilderness near the Canadian border. In March 1982 they voted to take strong action against the acid-forming pollution that is blown into the skies of Minnesota from other states. The legislature passed a law requiring the Minnesota Pollution Control Agency to prepare a plan by 1986 that controls the amount of sulfuric acid in the precipitation falling on the state. The law also requires the agency to adopt air quality standards for acid-forming sulfur dioxide emissions from Minnesota.

For the first time, then, a state government has gone beyond the conventional pollution control standards, which deal only with the amount of pollutants in the air around a particular pollution source at a given time. Minnesota officials hope that the new law will serve as a model for other states and for the U.S. government. It may also serve as a legal basis for Minnesota officials to argue in the federal courts that states such as Ohio which release excessive quantities of acid-forming pollutants are lawbreakers and should be forced to reduce their sulfur dioxide emissions. Such legal action, state officials believe, may be the only way to save the Boundary Waters wilderness.

In July 1982, the Senate Environment Committee became the first congressional body to go on record in favor of an active acid rain control program. In defiance of the Reagan administration, the fifteen members of the Senate committee voted unanimously to require a substantial reduction of sulfur dioxide emissions from power plants in thirty-one eastern states. The committee's twelve-year program would reduce U.S. sulfur dioxide emissions by about 50 percent at a cost to electric customers of $3.3 billion to $3.9 billion a year. Most of the cost, which would begin appearing in electric bills in the early 1990s, would be paid for by the midwestern states that have the largest sources of sulfur dioxide emissions.

Senator George Mitchell of Maine proposed the program as an amendment to the U.S. Clean Air Act, which is under review by Congress. Mitchell's amendment would establish a thirty-one-state acid rain control region extending from the Mississippi

River Valley to the East Coast. Each state in this region would be required to reduce its sulfur dioxide emissions by an amount to be determined by its current level of emissions. Depending on the emissions control technology used, states would have until 1992 or 1995 to comply with the law.

The Mitchell amendment will not be considered by the full Senate until work is completed on the Clean Air Act bill. The unanimous vote in the Senate Environment Committee indicates that the acid rain control program has strong bipartisan support. But given the opposition of the Reagan administration and the Edison Electric Institute, which represents electric utilities, approval of the program by both houses of Congress is uncertain at best. And if it is approved, a presidential veto could undo all the efforts of concerned legislators.

Clearly, the 1980s will be a crucial decade in the struggle over acid rain. Americans, Canadians, Europeans, and, indeed, all the peoples who breathe the air of the Earth and drink from its waters have a great deal at stake in the outcome. Will we leave a lethal legacy for future generations? Or will we act to preserve the beauty, integrity, and economic value of our natural heritage? The errors of the past have been duly recorded. The future is an open book. What we write in it will stand as our legacy for the generations of life on earth to come.

Notes

Chapter 1

1. Isaiah, 55:10.

2. Koran, Chap. 16.

3. Gene E. Likens, et al, "Acid Rain," *Scientific American*, October 1979, p. 47; Holum, John R, *Topics and Terms in Environmental Problems* (New York: John Wiley and Sons, 1977), p. 619.

4. U.S. Environmental Protection Agency National Air Pollution Emission Estimates, 1970-78 (EPA-450/4-80-002), 1980.

5. Canada, Ontario Ministry of Environment, "The Case Against the Rain," October 1980, p. 9.

6. Likens et al, "Acid Rain," p. 47.

7. C.K. Graves, "Rain of Troubles," *Science '80*, July/August 1980, p. 77.

8. Likens et al, "Acid Rain," p. 45.

9. Robert Trumbule, "Acid Precipitation: A Serious and Growing Environmental Problem," Library of Congress Report, February 13, 1981.

10. Charles V. Cogbill, "The History and Character of Acid Precipitation in Eastern North America," Proceedings of the First International Symposium on Acid Precipitation and the Forest Ecosystem, USDA General Technical Report NE-23, 1976, pp. 365-366.

11. Likens et al, "Acid Rain," p. 45.

12. Associated Press (from direct wire copy), "15 States 'Extremely Vulnerable' to Acid Rain," October 6, 1981.

13. Norway, Summary Report of the SNSF Project of the Agricultural Research Council of Norway, "Impact of Acid Precipitation in Forest and Freshwater Eco-

systems in Norway," Oslo, Norway, 1976, p. 96.

14. Anne La Bastille, "Adirondack Park Agency Report on Acid Precipitation," July 1979.

15. Canada, Ontario Ministry of Environment, "The Case Against the Rain," p. 2.

16. *Minneapolis Star,* "Study Says Acid Rain Threat Worse," December 11, 1981.

17. Gene Likens, interview with author, October 1979.

18. Canada, House of Commons, "Still Waters: Report by the Subcommittee on Acid Rain," 1981, Executive Summary, p. 2.

19. Peter Brimblecombe, "Attitudes and Responses Towards Air Pollution in Medieval England," *Journal of the Air Pollution Control Association,* October 1976, p. 944.

20. Ibid, p. 943.

21. Peter Brimblecombe, "London Air Pollution, 1500-1900," *Atmospheric Environment,* 1977, p. 1159.

22. Brimblecombe, "Attitudes and Responses Towards Air Pollution," p. 945.

23. National Academy of Science, "Acid From the Sky," *Mosaic,* July/August 1979, p. 36.

24. *Minneapolis Star,* "Acid Rain Threat to BWCA," January 25, 1979. (Article was published before findings were officially released.)

25. *Minneapolis Star,* "Threat From Acid Rain Worsens," July 11, 1979.

26. David R. Brower, "As Right As Acid Rain," *Not Man Apart,* December 1979, p. 6.

27. *Wall Street Journal,* "Acid Rain," June 30, 1980.

28. Associated Press (untitled story from direct wire copy), June 5, 1980.

29. Associated Press (from direct wire copy), "Bill Forces Some Power Plants to Shift from Oil to Coal," June 25, 1980.

30. *New York Times*, "Effort to Soften Environmental Rules Likely," November 18, 1980.

31. Gregory S. Wetstone, "The Need for a New Regulatory Approach," *Environment*, June 1980.

Chapter 2

1. Vincent Engels, "At the Sources of the Hudson," *The Conservationist*, May/June 1978, p. 13.

2. *New York Times*, "Uncertainty Voiced over Acid Rain," September 15, 1981.

3. James Gannon, "Acid Rain Fallout," *National Parks and Conservation Magazine*, October 1978, p. 19.

4. Carl Schofield, interview with author, October 1979.

5. Norway, Summary Report of the SNSF Project of the Agricultural Research Council of Norway, "Impact of Acid Precipitation on Forest and Freshwater Ecosystems in Norway," Oslo, Norway, 1976, p. 104.

6. Carl Schofield, interview with author, October 1979.

7. Government of Canada Freshwater Institute, "Acid Rain," 1980, p. 6.

8. "Aluminum Pollution Caused by Acid Rain Killing Fish in Adirondack Lakes," *BioScience*, July 1978, p. 472.

9. George Hendrey, interview with author, October 1979.

10. Ibid.

11. F. Harvey Plough, "Acid Precipitation and Embryonic Mortality of Spotted Salamanders," *Science.* April 2, 1976, p. 70.

12. David Johnson, interview with author, October 1979.

13. George Hendrey, testimony before International Joint Commission, Detroit, July 1979.

14. *Minneapolis Tribune,* "U.S. Study Says Sulfur Dioxide, Even Within Legal Limits, Can Harm Soybeans," January 8, 1981.

15. *Washington Post,* "Acid Rain Doesn't Hurt Most Crops, Actually Helps Some," April 1, 1980.

16. Robert H. Boyle, "An American Tragedy," *Sports Illustrated,* October 1981, p. 70.

17. *Cleveland Plain Dealer,* "Acid Rain: Hazardous to Your Health?" August 2, 1981.

18. *New York Times,* "Pollution Assessed in Ohio River Valley," March 1, 1981.

19. U.S. Congress, House hearings, Subcommittee on Oversight and Investigations, February 27, 1980.

20. *New York Times,* "Acid Rain Costs Money," April 26, 1981.

21. John R. Holum, *Topic and Terms in Environmental Problems* (New York: John Wiley and Sons, 1977), p. 5.

22. George Hendrey, interview with author, October 1979.

23. *Los Angeles Times,* "Acid Rain Poses Potential Danger," May 15, 1980.

24. *New York Times,* "Urban Pollution is Turning Glory That Was Rome to Dust," March 16, 1980.

25. *New York Times*, "Inspectors Find Statue is Corroded," May 19, 1980.

26. *New York Times*, "Greece Cuts Pollution to Save Acropolis," April 16, 1980.

Chapter 3

1. Gene E. Likens et al, "Acid Rain," *Scientific American*, October 1979, p. 47.

2. U.S. Environmental Protection Agency Draft Decision Series Report, "Acid Rain," 1979, p. 32.

3. Gene E. Likens, "Acid Precipitation," *Chemical and Engineering News*, November 22, 1976, p. 37.

4. U.S. Environmental Protection Agency Draft Report—Acid Rain, p. 32.

5. Robert H. Boyle, "An American Tragedy," *Sports Illustrated*, October 1981, p. 70.

6. Likens et al, "Acid Rain," *Scientific American*, p. 45.

7. Earl Finbar Murphy, "The More Things Change," Proceedings of the First International Symposium on Acid Precipitation and the Forest Ecosystem, USDA General Technical Report NE-23, 1976, p. 44.

8. Likens et al, "Acid Rain," *Scientific American*, p. 3.

9. Richard A. Kerr, "Global Pollution: Is the Arctic Haze Really Industrial Smog?" *Science*, July 20, 1979, p. 293.

10. Ibid, p. 290.

11. U.S. Environmental Protection Agency, "Research Summary: Acid Rain," (EPA 600/8-79-028), October 1979, p. 4.

12. George T. Wolff et al, "Acid Precipitation in the New York Metropolitan Area." *Environmental Science and Technology*, February 1979, p. 209.

13. Robert Trumbule, "Acid Precipitation: A Serious and Growing Environmental Problem," Library of Congress Report, February 13, 1981.

14. Charles V. Cogbill, "The History and Character of Acid Precipitation in Eastern North America," Proceedings of the First International Symposium on Acid Precipitation and the Forest Ecosystem, USDA General Technical Report NE-23, 1976, p. 366.

15. Gene E. Likens et al, "Acid Rain," *Environment*, March 1972, p. 36.

16. Cogbill, "History and Character of Acid Precipitation," p. 369.

17. Likens et al, "Acid Rain," *Environment*, p. 37.

18. *Los Angeles Times*, "Acid Rain Called Southland Problem," April 21, 1979.

19. Ibid.

20. Ibid.

21. U.S.-Canada Research Consultation Group on the Long Range Transport of Air Pollutants, "The LTRAP Problem in North America: A Preliminary Overview," 1979, p. 13.

22. U.S.-Canada Research Group, "Second Report," November 1980, p. 19.

23. "The Not So Gentle Rain," *Canada Today/ D'Aujourd'Hui*, February 1981, p. 7.

24. U.S.-Canada Research Group, "The LTRAP Problem," pp. 11-13.

25. Likens et al, "Acid Rain," *Scientific American*, p. 44.

26. William M. Lewis, Jr., and Michael C. Grant, "Acid Precipitation in the Western United States," *Science*, January 11, 1980, p. 176.

27. Ibid.

28. Ibid. p. 177.

29. Erhard M. Winkler, "Natural Dust and Acid Rain," Proceedings of the First International Symposium on Acid Precipitation and the Forest Ecosystem. USDA General Technical Report NE-23, 1976, p. 209.

30. Eville Gorham, interview with author, November 1980.

31. Likens et al, "Acid Rain," *Scientific American*, p. 44.

32. U.S. Environmental Protection Agency Report (EPA-450/1-78-003), 1978.

33. Gus Speth, "The Sisyphus Syndrome: Acid Rain and Public Responsibility," *National Parks and Recreation*, February 1980, p. 13.

34. Frank Corrado, "Midwest in the Soup; Hazy Blobs Abound," *EPA Environment Midwest*, July 1979, p. 12.

35. Kurt H. Hohenemser, "Energy: Downwind," *Environment*, April 1977, p. 2.

36. Corrado, "Midwest in the Soup," p. 16.

Chapter 4

1. Svante Oden, interview with author, January 1982.

2. Norway, Summary Report of the SNSF Project of the Agricultural Research Council of Norway, "Impact of Acid Precipitation on Forest and Freshwater Ecosystems in Norway," Oslo, Norway, 1976, p. 88.

3. Ibid.

4. Kenneth Hood, interview with author, May 1980.

5. Svante Oden, "The Acidity Problem—An Outline of Concepts." Proceedings of the First International

Symposium on Acid Precipitation and the Forest
Ecosystem, USDA General Technical Report NE-23,
1976, p. 7.

6. Ibid., p. 8.

7. Ibid., p. 24.

8. Ibid., p. 4.

9. Ibid., p. 12.

10. Ibid., p. 14.

11. John R. Holum, *Topics and Terms in Environmental Problems* (New York: John Wiley and Sons, 1977),
p. 5.

12. SNFS Project, "Impact of Acid Precipitation,"
p. 8.

13. Anne La Bastille, "Acid Rain: How Great a
Menace?" *National Geographic*, November 1981,
p. 653.

14. SNFS Project, "Impact of Acid Precipitation."
p. 11.

15. Oden, "The Acidity Problem," p. 21.

16. Eville Gorham, interview with author, November
1979.

17. Oden, "The Acidity Problem," pp. 26-27.

18. SNFS Project, "Impact of Acid Precipitation,"
p. 104.

19. Ibid.

20. Ibid.

21. Ibid.

22. Ibid.

23. Oden, "The Acidity Problem," p. 1.

24. Ibid., p. 33.

25. Svante Oden, interview with author, January
1982.

Chapter 5

All quotations and other material attributed to the scientists profiled in this chapter are based on interviews conducted by the author between 1979 and 1982.

Chapter 6

1. Robert Rauch, interview with author, May 1980.
2. *Minneapolis Star*, "United States Cuts Oil Imports 18% in 1980," January 16, 1981.
3. General Electric Co., *U.S. Energy Data Book 1980*, p. 7.
4. 1980 Congressional Quarterly Almanac, p. 507.
5. Sierra Club National News Report, "Oil Conservation Plan Unveiled," March 14, 1980.
6. U.S. Congress, House hearings, Subcommittee on Oversight and Investigations, "Acid Rain," February 27-28, 1980, p. 370.
7. Ibid., p. 371.
8. Ibid.
9. Ibid., p. 372.
10. Ibid.
11. U.S. Congress, Senate hearings, Subcommittee on Environmental Pollution, p. 18.
12. *Cleveland Plain Dealer*, "Canada Sees Acid Rain Kill Life in Lakes," August 2, 1981.
13. Ibid.
14. *Cleveland Plain Dealer*, "Snarls on the Subject," August 2, 1981.
15. Peter Towe, interview with author, May 1980.
16. Raymond M. Robinson, "Acid Rain, Canada, and the Coal Conversion Program," *Journal of the Air Pollution Control Association*, May 1980.

17. George Rejhon, interview with author, May 1980.

18. Robert Rauch, interview with author, May 1980.

19. Senate hearings, Subcommittee on Environmental Pollution, pp. 184-85.

20. *New York Times*, "Wrong Place For Coal Billions," March 12, 1980.

21. Senate hearings, Subcommittee on Environmental Pollution, p. 233.

22. *New York Times*, "We Don't Need Government Bribes," March 15, 1980.

23. House hearings, Subcommittee on Oversight and Investigations, p. 244.

24. Ibid., p. 365.

25. Ibid., p. 277.

26. Memorandum from House Subcommittee on Oversight and Investigations Staff to the Honorable Bob Eckhardt, chairman, February 26, 1980, "Exhibit for February 28 Hearing."

27. House hearings, Subcommittee on Oversight and Investigations, p. 296.

28. Ibid.

29. Rafe Pomerance, interview with author.

30. 1980 Congressional Quarterly Almanac, p. 508.

31. Ibid.

Chapter 7

1. Robert Rauch, interview with author, May 1980.

2. C.K. Graves, "Rain of Troubles," *Science 80*, July/August 1980, p. 79.

3. Carl Bagge, remarks to Washington, D.C., Journalism Center Seminar, September 1979.

4. U.S. Congress, House hearings, Subcommittee on Oversight and Investigations, "Acid Rain," February 26-27, 1980, p. 46.

5. *Washington Post*, "EPA Official Raps Schlesinger on Coal Burning," June 22, 1979.

6. House hearings, Subcommittee on Oversight and Investigations, p. 194.

7. Proceedings of the U.S. Environmental Protection Agency Symposium on Flue Gas Desulfurization. (EPA-699/9-81-019A), p. 120.

8. Electric Power Research Institute Technical Assessment Guide, July 1979.

9. *Cleveland Plain Dealer*, "Scrubbers Do The Job But Cost Is High," August 7, 1981.

10. Ibid.

11. *Minneapolis Star*, "Bacteria Tamed To Fight Pollution and Help Industry," October 27, 1981.

12. David R. Brower, "As Right As Acid Rain," *Not Man Apart*, December 1979, p. 6.

13. *Los Angeles Times*, "Acid Rain Falls on Canada But Blame Falls on U.S.," October 19, 1981.

14. House Hearings, Subcommittee on Oversight and Investigations, p. 398.

15. Ibid., pp. 42-47.

16. Ibid., p. 47.

17. National Coal Association, "Some Facts About Acid Rain," 1980, p. 15.

18. Ibid.

19. George Hendrey, interview with author, October 1979.

20. House Hearings, Subcommittee on Oversight and Investigations, pp. 193-194.

21. National Public Radio. "All Things Considered." August 13, 1979.

22. George Hendrey, interview with author, October 1979.

23. House Hearings, Subcommittee on Oversight and Investigations, pp. 28-29.

24. Eville Gorham, interview with author, October 1979.

25. "Acid Rain Fallout: Pollution and Politics," *National Parks and Conservation*, October 1978, p. 17.

26. George Hendrey, interview with author, October 1979.

27. *Los Angeles Times*, "Acid Rain Falls on Canada."

28. Canada, House of Commons, "Still Waters: Report by the Subcommittee on Acid Rain," 1981.

29. *Los Angeles Times*, "Acid Rain Falls on Canada."

30. Ibid.

31. President Jimmy Carter, White House Message on the Environment, August 2, 1979.

32. *Washington Post*, "Onions, Tears and Rain," August 13, 1979.

33. Brower, "As Right as Acid Rain," p. 6.

34. Robert Rauch, interview with author, May 1980.

35. House hearings, Subcommittee on Oversight and Investigations, p. 412.

36. *Cleveland Plain Dealer*, "Scrubbers Do the Job."

37. David Johnson, interview with author, October 1979.

38. *St. Paul Pioneer Press*, "Acid Rain No Problem, Says Edwards," March 7, 1982.

39. Robert H. Boyle, "An American Tragedy," *Sports Illustrated*, October 1981, p. 82.

40. Earl Finbar Murphy, "The More Things Change."

Proceedings of the First International Symposium on Acid Precipitation and the Forest Ecosystem, 1976, p. 45.

41. Ibid., p. 46.

42. Ibid., p. 45.

43. *Minneapolis Tribune*, "Five Protestors Leave Smokestacks," February 12, 1982.

.

Bibliographical Essay

General Reference on Acid Rain

Perhaps the most useful, comprehensive reference on acid rain is the record of two days of hearings held before a subcommittee of the U.S. House of Representatives in 1980. The document, "Acid Rain, Testimony before the Subcommittee on Oversight and Investigations, February 27-28, 1980," includes presentations by expert witnesses on all aspects of the acid rain problem: the perspectives of the electric power industry, scientists, and environmentalists; information on the effects of acid precipitation; discussions of possible means of mitigating acid-caused damage; and discussions of the political, social, and economic ramifications of the acid rain threat.

Another general reference work is entitled "Proceedings of the First International Symposium on Acid Precipitation and the Forest Ecosystem," USDA General Technical Report NE-23, 1976. This document provides a global perspective on acid rain, as well as detailed information on the causes and effects of the phenomenon.

A shorter, but still fairly comprehensive and authoritative, appraisal of the acid rain problem can be found in an article entitled "Acid Rain" in *Scientific American*, October 1979, by Gene E. Likens of Cornell University and several colleagues. Likens also was the principal author of an earlier article entitled "Acid Rain" in the March 1972 issue of *Environment* that is regarded as the first major explanation of acid rain in North America.

Canada and North America

A report by the Canadian House of Commons Subcommittee on Acid Rain, "Still Waters," is an excellent overview of the acid rain problem in Canada. More detailed information on Canadian acid rain and its relationship to U.S. sources of pollution can be found in a pair of reports by the U.S.-Canada Research Consultation Group on the Long-Range Transport of Air Pollutants. The reports are called "The LTRAP Problem in North America: A Preliminary Overview, 1979" and "The LTRAP Problem in North America: Second Report, 1980."

Scandinavia

Among the best sources of information about the acid rain problem in Sweden and about the history of acid rain research there are: an article by Swedish acid rain researcher Svante Oden beginning on page 7 of the above-mentioned "Proceedings of the First International Symposium on Acid Precipitation and the Forest Ecosystem" entitled "The Acidity Problem—An Outline of Concepts;" and an article in the November 1981 issue of *National Geographic* by Anne La Bastille, "Acid Rain: How Great a Menace?"

An excellent reference on acid rain in Norway is "Precipitation in Forest and Freshwater Ecosystems in Norway" from The Summary Report of the SNSF Project of the Agricultural Research Council of Norway (Oslo, Norway, 1976).

Index